In the Search for Beauty

Unravelling non-Euclidean Geometry

Search
In the
for Beauty

Unravelling non-Euclidean Geometry

Voldemar Smilga

World Scientific

NEW JERSEY · LONDON · SINGAPORE · BEIJING · SHANGHAI · HONG KONG · TAIPEI · CHENNAI · TOKYO

Published by

World Scientific Publishing Co. Pte. Ltd.

5 Toh Tuck Link, Singapore 596224

USA office: 27 Warren Street, Suite 401-402, Hackensack, NJ 07601

UK office: 57 Shelton Street, Covent Garden, London WC2H 9HE

British Library Cataloguing-in-Publication Data
A catalogue record for this book is available from the British Library.

First published 2019 (Hardcover)
Reprinted 2019 (in paperback edition)
ISBN 978-981-120-830-0 (pbk)

IN THE SEARCH FOR BEAUTY
Unravelling Non-Euclidean Geometry

ISBN 978-981-3274-35-8

For any available supplementary material, please visit
https://www.worldscientific.com/worldscibooks/10.1142/11102#t=suppl

Printed in Singapore

Foreword

Voldemar Smilga (1929–2009) was a physicist. Since 1963 and till the end of his life he worked in Kurchatov's Institute in Moscow. His most known scientific achievement is the invention of the so-called μSR (muon spin rotation) method to study microscopic magnetic fields in different media.[1] But besides writing scientific papers and specialised monographs, V. Smilga wrote three popular books addressed to everyone interested in science and in the history of science.

The first book, *Relativity and Men*, tackles the relativity theory and the history of its creation. The last book, *Molodye gody Nikolaya Ivanovicha Lobachevskogo*, published in Russian, after his death, is an essay on the history of Kazan, of the Kazan University and describes the youth of Nikolai Lobachevsky, who studied and worked there. And we now recommend to the reader his second book devoted to the history of attempts to prove the fifth Euclidean postulate, which finally led to the discovery of non-Euclidean geometry.

For two thousand years since the time when Euclid's *Elements* were published, mathematicians tried to prove the fifth postulate (and did not try to prove the four others) because it was formulated by Euclid in a complicated, one can even say ugly way. Indeed, the original Euclid's formulation was: *If a straight line crossing two straight lines makes the interior angles on the same side that sum to less than two right angles, then the two lines,*

---

[1]The positive muon μ^+ is an elementary particle, a heavy analog of the positron. When introduced in matter, it feels microscopic magnetic fields which rotate its spin (the angular momentum). Finally it decays emitting positrons, and it appears that these positrons mostly go along the direction of the muon spin at the moment of its decay. Knowing the initial muon spin direction and the direction at the moment of the muon decay, one can extract information about the magnetic fields that acted on the muon in between.

if extended indefinitely, meet on that side on which the sum of the interior angles is less than two right angles.

It offended the aesthetic feelings of the scholars. They were convinced that a basic axiom, from a set of which all geometric properties are derived, could not be so unsightly. Searching for beauty, the mathematicians from Proclus to Omar Khayyam (who was not only a brilliant poet, but also a brilliant mathematician) to Gauss tried again and again to prove it as a theorem. But all these attempts failed.

Finally, in the 19th century, Gauss, Lobachevsky and Bolyai guessed and then Klein proved that such a proof cannot be constructed in principle. The fifth postulate (its simpler formulation, the way it is now studied at schools reads: *Given any straight line and a point not on it, there exists on the plane containing this line and this point one and only one straight line which passes through that point and never intersects the first line, no matter how far they are extended.*) can be accepted as an axiom and this leads to Euclidean geometry. But one can also accept than one *cannot* draw such a non-intersecting line—this is true in the geometry describing Earth's surface—or insist that one can draw two different such lines, and this leads to Lobachevsky's geometry.

In the 20th century, Einstein understood that the true geometry of our world is non-Euclidean.

When my father was writing this book, I was about 10 or 11. I was interested in geometry and read the manuscript. And I was a little disappointed by the fact that many pages of the book did not involve mathematical statements, but told stories about *mathematicians* rather than about mathematics. I complained about this to my father (human stories were not so interesting for me at that time), but he ignored my complaint. As a result, this book, the way it is written, might be interesting not only to science-lovers, but also to people having humanitarian minds. A reader may *skip* all mathematical details, if s/he wishes (though, of course, paying attention to these details would add value to the reader's comprehension and appreciation) and concentrate on the dramatic personal stories of the scholars who sought for elusive mathematical beauty for many centuries, and when the beauty was eventually found, it was another kind of beauty, not the same that the people expected to find.

This book was first published in Russian by *Molodaya Gvardiya* ("Young Guard") in Moscow in 1965 and then republished in 1968. The witty and amusing illustrations were provided by G. Kovanov and V. Kovynev. In 1970 the book was translated into English by George Yankovsky and published by *Mir*. These Soviet-time publishers do not exist any more, and the original editions have become bibliographic rarities. The Russian version of the book was republished in 2006 by *Moskovskie Uchebniki* and now *World Scientific* presents to you the new English edition. For the new edition, I've edited a little of Yankovsky's translation. Firstly, Yankovsky did not translate for some reason certain portions of the text, and I have translated them myself. Mostly, it was scattered individual phrases and paragraphs, but in Chapter 7 about a couple of pages of the original text were left out.[2] Secondly, I've removed some bugs—mostly, in mathematics (the terminology and sometimes the essence) and the English spellings of Arabic names. Thirdly, I've modified Yankovsky's choice for the English translations of Omar Khayyam's verses (see the footnote on p. 73).

Andrei Smilga

[2]As a small puzzle, I propose to the reader to try to find this place ☺.

Acknowledgments

It is a pleasure for me to thank my son Boris Smilga for the essential help with the editing. I also would like to thank the whole World Scientific team, and especially Ms. L. Narayanan for her encouragement and support.

Contents

Chapter 1

Before Euclid—Prehistoric Times

The true beginning of this story goes back to times immemorial.

Where was it, when and how did geometry come into being? Where, how and when did it take shape and become a science? Who was the very first to propose the axiomatic structure of geometry?

We do not know, and most likely never will.

It is generally believed that he was a Greek. But perhaps the glorified priests of Egypt or the renowned chaldean magi are the true fathers of science.

However all that may be, geometry arrived in Greece in the seventh century before the Christian era.

It was there and then that the Greeks, admirers of cold logic and the exquisite elegance of pure intellect, lovingly polished to a brilliance (or perhaps originated) one of the most beautiful creations of human thought—geometry.

Elegance indeed, yet actually the matter was far more involved and intriguing. One thing is certain, and that is that geometry sprang from practical needs.

The development of logic (and consequently geometry as well) was influenced to some extent by the Greeks' devotion to law and oratory. But in Egypt, too, geometry was important to men of the practical world—very important. And as for endless litigations and court proceedings, the Greeks were far behind the country of the pharaohs.

In a word then, a serious analysis of this question would take us too far astray; let us be satisfied with the fact. Geometry has established itself. This is the start of a gripping, dramatic contest in pure logic that has continued for two and a half thousand years.

The history of the fifth postulate goes back just about as many years.

It is as dramatic as it is instructive, a detective story with an unexpected but happy ending.

Now for the story.

Geometry, we believe, began with the Ionic school. To be more precise, its founder was Thales of Miletus, who was believed to have lived close to a hundred years (either from 640 to 540 or 640 to 546 BC).

We don't seem to know very much about him.

We know for sure he had the title of one of the Seven Sages of Greece; we also know that in accordance with the established reckoning he was the first philosopher, the first mathematician, the first astronomer and, generally, the first in all sciences in Greece. We might say that he was to the Greeks what Lomonosov was to the Russians: THE FIRST.

As a young man he most likely made his way to Egypt on affairs of trade, for he began his career as a merchant. Here, the pharaoh Psammetichus had just lifted the "iron curtain" and was beginning to allow foreigners into his country.

Thales remained in Egypt for a good number of years, studying in Thebes and Memphis. Later he returned to Greece and founded a school of philosophy. Obviously, he appeared more as a popularizer of Egyptian wisdom than as an independent thinker.

The view is that he brought with him geometry and astronomy.

At any rate, there is one thing that all philosophers can learn from him—and that is conciseness. Legend has it that his complete works (which naturally were all lost) consisted of only about 200 poems.

We can only conjecture what he accomplished in geometry, though Greek authors attributed a great deal to him.

For instance, Proclus Diadochus (we will be meeting him again) claims that it was Thales—no other—that proved the theorems that:

(a) vertical angles are equal;

(b) the angles at the base of an isosceles triangle are equal;

(0) a diameter divides a circle in half.

And some others.

Assuming even that the historians of science wrote the exact truth, we still do not know whether Thales himself arrived at these theorems or simply repeated ideas of the Egyptians.

Perhaps the only definitely established fact of the scholarship of Thales of Miletus is his prediction of the solar eclipse of 585 BC.

But legends grew up around him in hosts. And this in itself indicates that he was a scholar of stature.

One of the stories is particularly dear to men of learning. It is Aristotle who relates it:

"When Thales was reproached for his poverty, since, as they said, studies in philosophy do not create any profit, it is said that Thales, foreseeing a rich harvest of olives on the basis of astronomical findings, advanced during winter a small sum of money he had accumulated to the owners of all the oil-presses in Miletus and on the island of Chios. He was able to engage the oil-presses cheaply, for there was no competition from anywhere. When harvest time arrived, there was a sudden demand by many people for the oil-presses. Thales then rented out the oil-presses at prices that he himself desired.

"Thales thus accumulated a great deal of money and proved in this manner that it is not difficult for philosophers too to become rich, the only thing is, however, that that is not the subject of their interests."

We do not know what Thales did with the money he made in this successful practical application of astronomy. We hope he spent it as a true philosopher would.

His pupils and followers apparently paid proper attention to geometry in their philosophical deliberations. However, the central mathematical school of the 6th and 5th centuries BC was the Pythagorean school.

The authentic biographical information about Pythagoras boils down, in essence, to a few stories. In this respect, he is much like Thales of Miletus. The obscurities begin with his origin.

Bertrand Russel sums the matter up by saying that some believe him to be the son of a wealthy citizen named Mnesarch, others the son of the god Apollo, and adds that the reader can take his pick.

It is further believed that Pythagoras lived just as long a life as Thales— something in the vicinity of one hundred years (perhaps 569 to 470 BC).

Again like Thales, he spent some twenty years in Egypt imbibing wisdom, but later (here he surpassed Thales) he lived about ten years in Babylonia adding still more to his store of knowledge. It is also claimed that he travelled in India, but nobody seems to believe it.

Boxers claim that Pythagoras took boxing laurels in the Olympic games, but the source of such claims is never indicated. I have nothing to support them either. As in the case of Thales, the exciting thing is the unexpected combination of philosopher, mathematician and boxer.

Pythagoras may not have done much in boxing, but in politics he did, and very actively, though not at all successfully.

The citizens of the Sicilian town of Crotona, where he founded his school after his wanderings in distant lands and also got the town involved in an exhausting war, finally asked him to leave together with his school. Which he did in rather much of a hurry, which was a reasonable thing to do.

As a mathematician and scholar he was a giant, but nevertheless he does not call forth great admiration. His Pythagorean order of philosophers and mathematicians is much too reminiscent of a barracks and Pythagoras himself suspiciously resembles a führer, though much more cultured than any of those of the twentieth century.

It is precisely Pythagoras himself—most likely in a campaign to build up his authority—who built up and popularized the idea that his loving father was the fair-haired effulgent Apollo. Actually he became the true father of the presently popular custom of attributing to himself the scientific results of his pupils. There, the matter was quite official. There existed a fiat according to which the author of all the mathematical studies of the school was to be named Pythagoras.

Though one might repeat that such things are done right and left today, the passage of 25 centuries has greatly softened and civilized the customs. The essence is the same, but the form has become ennobled.

Pythagoras is the unsurpassed leader here because he handled matters so that his faithful pupils claimed him author of work done long after his death. Quite understandably then—that being the state of affairs in the Pythagorean school—that the most cogent of all arguments was a simple reference to The Authority Himself.

That is exactly how the wording went: "He said so Himself". After which any discussion was totally out of place—even dangerous.

He and his dear pupils also held in secret their methods for solving mathematical problems. Too, he compiled for the members of his order a long list of taboos.

I quote from the rules of good manners of the gentlemen of the Pythagorean Club:

"1. Restrain from using beans in your food.

"2. Do not pick up what has fallen.

"3. Do not touch white roosters.

"4. Do not take a bite from a whole bread.

"5. Do not walk on a highway.

"6. When removing a pot from the fire, do not leave traces in the ashes, but mix the ashes."

The list could be extended. It was this bunch that rose to power in one Greek town, then in another, implanting the cult of Pythagoras and, accordingly, demanding compliance with their statutes. With melancholy, Bertrand Russel relates that those who were not reborn in the new faith thirsted for beans and so sooner or later rose up in arms.

It is also told that he preached to the animals, for he made little distinction between them and human beings.

But the Pythagorean school advanced geometry and mathematics in general. Very much so, in fact. All of this taken together is not a bad illustration of the danger of idealizing representatives of the exact sciences and of the intellect generally.

Incidentally, to us, Pythagoras is mainly a mathematician. Yet he himself and his contemporaries took the view that his profession was that of a prophet. That was of course their business, they were closer to events. But, as we know, every prophet must be in part magician, demagogue and charlatan.

Pythagoras was apparently past master in each field. The pupils tried hard too. According to one story, one of his hips was of gold, to another that reliable people saw him at two different places at the same time, to a third that when he was wading across a stream, the water overflowed the banks crying "Long Live Pythagoras!"

True, the Greeks had a goodly number of reasonable people.

Xenophanes, the well-travelled, realistic, freethinking and malicious-tongued philosopher and writer, spoke of Pythagoras in a rather different vein. One of his epigrams went: Pythagoras witnessing a puppy being beaten said: "Do not hit him, it is the soul of a friend of mine. I recognized it when I heard it cry out."

The teaching of transmigration of souls is one of the basic elements in the overall conception of Pythagoras, and Xenophanes, as the reader can see, had a pointed thing or two to say on that score.

Heraclitus was very strict in his portrait of Pythagoras, "multiple knowledge without reason".

We leave Pythagoras, but before doing so, just one more curious story by one of his honoured admirers. How devious indeed are the pathways of science. Quite naturally, geometry, like all branches of knowledge, was most carefully concealed from the common people by the Pythagoreans. Who knows, perhaps to this day no one would know of geometry (outside the Pythagoreans) if it weren't for....

But here is the legend as to how the Pythagoreans account for the spread of geometry. One of them is to blame, for he lost the money of the community. After that calamity, the community permitted him to earn the money by teaching geometry, and geometry was given the name "the legend of Pythagoras".

A curious thing is that there seems to have been a geometry textbook by that name.

As to the story itself, if there is a grain of truth in it at all, then, though I do not consider myself a malicious person, I would be pleased to learn that the truant Pythagorean had not lost the money after all but had spent it in a spree in the local port tavern swilling wine, eating a white rooster with beans, biting a whole roll of white bread and singing drunken songs on the highway.

Another man contributed greatly to geometry, and again to my taste he was an unpleasant character.

His name was Plato (428 to 348 BC).

In his views, in his methods of setting up a school, and in his love of self-advertisement, Plato much resembles Pythagoras. But before I say why I do not like him, let me explain what his most significant contribution to geometry is.

He is considered—and perhaps justly so, for I am not a specialist in the field—one of the greatest philosophers of Greece. Indeed he did a great deal for the development of mathematics and valued it highly. At the entrance to his Academy he had, hewn in stone, the inscription: "Let no one destitute of geometry enter my doors!" The point is that Plato believed that "the study of mathematics brings us closer to the immortal gods", and educated his pupils in this spirit, adding mathematics where it was needed and where it wasn't. Some of his pupils became brilliant geometers. Plato had numerous pupils and they naturally spread numerous stories praising the teacher.

It was apparently Plato who first made the explicit demand that mathematics generally and geometry in particular be constructed in deductive fashion. To put it differently, all the propositions (theorems) must be rigorously logically deduced from a small number of basic statements called axioms.

This was a momentous step forward.

By the time Plato arrived on the scene, geometry had developed extensively.

A multitude of extremely complicated problems had been solved and highly involved theorems proved. What was apparently lacking was a clear-cut general scheme of construction. As is frequently the case in science, the development of geometry was spurred tremendously by three problems that adamantly refused to succumb.

Since we have gone this far, I will state the problems:

It was required, with the aid of compass and straightedge alone (no other instruments allowed), to

(1) divide a given angle into three equal parts (trisecting the angle);

(2) construct a square of area equal to the area of a given circle (squaring the circle);

(3) construct a cube of volume twice that of a given cube (duplicating the cube, the "Delphian problem").

It was only at the end of the nineteenth century that it was proved that, thus posed, not one of the problems is solvable, though all three are readily resolvable if other geometrical instruments are employed. To put it in other words, the problems can be handled by utilizing loci different from a straight line or an arc of a circle.

But the Greek rules only permitted compass and straightedge.

Plato even substantiated this requirement by some sort of reference to the authority of the gods.

That is why neither of there problems was solved, but in the effort geometry was greatly developed.

Too bad we have no place or time for the numerous exciting stories that go along with these problems. But we will recall a legend to show that we are objective in our attitude towards Plato. One of the versions of this story makes him out a very reasonable man.

Eratosthenes relates that once, on the island of Delos an epidemic of plague broke out. The inhabitants of the island naturally turned to the Delphian oracle who ordered to duplicate the volume of the golden cubical sacrificial repository to Apollo without altering its shape. Plato was asked to advise.

He did not resolve the problem but interpreted the oracle as meaning to say that the gods were angry with the Greeks for the endless internecine wars and desired that the Greeks should give up warfare and engage in the sciences, particularly geometry. The plague would then vanish.

Legends or no legends, Plato as philosopher and man is in my opinion extremely unpleasant. It is not even the fact that he was supporter of the most rabid idealism and on every occasion appealed to the gods. What is worse, he built up a theory of the state taking as his model nearby Sparta— a real haven of fascism. Too, the basic planks of his utopia fully conform to the demands of nazism.

He spent his whole life fighting tooth and nail against democracy in political life and against materialism in spiritual life.

He not only scourged the materialist-thinking philosophers abstractly in his philosophical writings, but, demonstrating a very practical approach to matters, often employed political denunciation—a beloved weapon in all ages to fight scientific opponents.

There is even a story that he bought up the works of his bitterest enemy Democritus so as to destroy them.

Democritus is a special topic of discussion.

If one agrees that the source of our modern physics is to be sought among the Greeks (and that is most likely the case), then the distance covered is great indeed—something like two thousand years from Aristotle to Newton. The four primal elements of Aristotle—air, water, earth and fire—marked one of the first attempts to define the concept of the "elementary particles" of physics.

True, the Greeks did not know physics in the modern sense of the word. At the heart of matters were speculative arguments, not experiment. But this is not so important to us now.

Perhaps it is the almost total absence of experiment that brings out the utterly amazing conjecture of the sly philosopher Democritus of Abdera.

Roughly half a century before Aristotle, he believed that all substances consisted of minute indivisible particles—atoms—and that the different properties of substances were determined by the different qualities of the atoms themselves. In a given substance, however, all atoms were identical and devoid of any individuality.

These views are so close in spirit to modern conceptions that one of the founders of quantum mechanics, Erwin Schrödinger, took great pleasure in startling his listeners with the elegant paradox: "The first quantum physicist was not Max Planck but Democritus of Abdera."

Most likely, Democritus would have been most amazed to hear this flattering comment, yet one must agree that Schrödinger surely has certain rights when it comes to discussing quantum theory.

The fate of Democritus' views is remarkable in yet another two ways. Firstly, not a single one of his writings has come down to us. Either Plato indeed succeeded in his neat little methods of scientific discussion, or simply the books were lost through the ages; at any rate, to our misfortune, the ideas of one of the first materialists in the world can be judged only on the basis of extracts and later retellings.

Secondly, the first popular-science treatise (and popularizers of science should never forget this) was devoted to a discussion of his ideas.

What is more, the book in question set a world record, for the poem is of extreme length. I am of course alluding to the poem on the Nature of Things (*De rerum natura*) by Titus Lucretius Carus, which was written some three hundred years after the death of Democritus—two thousand years ago.

By the way, Democritus had it rather good nevertheless, because traces of many other scholars (particularly among the materialists) have been lost completely. For instance, there is still great doubt about whether Democritus' teacher Leucippus ever lived. Then of course it is entirely conjecture whether Leucippus was co-author or author of the ideas of atomism.

There is also a version that the teaching of Democritus was borrowed from some chaldean magi granted to his father by the Persian king Xerxes.

And if we may permit ourselves a bit of moralizing, it is worth noting that in science ideas are incomparably longer-lived than the memory of those who engender them. Incidentally, most scientists in any branch of knowledge can grasp almost anything except this not-too-unexpected idea.

But whoever was the founder of the atomistic theory, and whether quantum mechanics has its source in Democritus or the chaldeans, the views of the atomistic school are roughly as follows.

The world consists of atoms and void. The atoms are indivisible. They are elementary and keep their properties in all conditions. Atoms do not succumb to any kind of outside action whatsoever, they are not generated and they are indestructible. Primordial distinctions exist between atoms, and these distinctions determine the variety of properties of all things.

What we today regard as elementary particles represents entities that are nor so similar to the atoms of Democritus. They appear and disappear, they convert from one into another, and they are readily acted upon—in a word, we must say that the Greeks were much more logical in their concept of an elementary particle than are the physicists of the twentieth century.

There is a reliable statement made by Archimedes which strongly suggests that Democritus was a marvellous geometer. It would seem that he was the one who computed the volume of a cone and a pyramid. That was a brilliant achievement, but unfortunately not many details are known. Be that as it may, of the forerunners of the integral calculus the first was apparently Democritus.

Another complicating circumstance is the fact that practically our only source is a book by Proclus Diadochus. Since Proclus was a follower of Plato, he hardly made any mention of his scholarly opponents.

Quite naturally, Democritus was enemy number one and was first to be banished from history.

The picture is practically the same with regard to Anaxagoras. We know hardly anything about the geometrical studies of that remarkable philosopher who was one of the first materialists. The only thing we do know is that in the dungeon where he resided because of his views, he investigated the problem of the squaring of the circle. His philosophical views definitely merit a good word.

Incidentally, this was best done by Plato. In one of his works we find a dialogue between an Athenian (Plato himself) and a Spartan. This is how Plato handles Anaxagoras.

Athenian: "When we seek to obtain proof of the existence of the gods and refer to the Sun, the Moon and the stars and the Earth as divine creatures, the pupils of these new sages object that all these things are only the ground and the stones and they (the stones, that is) are quite unable to take care of the affairs of human beings."

Obvious, then, that Anaxagoras and his pupils are simply a product of murky Tartarus.

The Spartan straightway perceives the heresy and cries out with indignation: "What awful harm is this for the family and for the state that flows from such attitudes of the young people!"

That is the way Plato carried out his discussions.

I would be very pleased if his contributions to the development of geometry turned out to be greatly exaggerated.

But as things stand today we must admit that his school brought forth a galaxy of brilliant mathematicians, and his is the first mention of the axiomatic method.

To summarize then, in the fourth and third centuries BC geometry was a fully developed science. With traditions, with fully elaborated methods of solving problems, mighty achievements and even a number of textbooks and schools of thought.

There is no need or place to go into the story of all the geometers of the pre-Euclidean period.

Suffice it to give a list of the mathematical giants of that period that preceded Euclid—Thales of Miletus, Anaximander, Amerist, Mandriat, Eonipidus, Anaximedes, Democritus, Anaxagoras, Pythagoras, Hippias, Archytas, Hippocrates of Chios (no relation to the physician), Antiphon,

Plato, Theaetetus, Eudoxus of Cnidus (these last two are towering figures, especially Eudoxus, who lived between 400 and 337 BC and is believed to have also been an astronomer, physician, orator, philosopher and geographer).

Menaechmus, Leodatus, Deinostratus, Aristaeus, Eudemus, Theophrastus, Theudius and yet another couple of dozen names.

And also Aristotle.

Aristotle, beyond any doubt, is one of the greatest minds in the history of humankind.

True, when balanced, the harm done by his works almost tips the scales against the good. Aristotle is hardly at all to blame for this, but in the Middle Ages, his works, pared down and purified to the point where they could no longer engender fresh thinking, became the principal weapon of reaction.

But an appraisal of his works is a whole history in itself. The only thing that need be said here is that he was definitely and deeply interested in geometry. Note that he paid special attention to the theory of parallel lines.

What is more, he contributed two extremely important propositions to this field. True, they do not appear in the works that have come down to us, but all succeeding mathematicians unanimously attribute these statements to Aristotle.

Jumping out ahead of our story, we may note that the cleverest proofs of Euclid's fifth postulate are based on "Aristotle's principle". We will come back later to what Aristotle said about the properties of parallel lines. In the meantime ...

Chapter 2

Euclid

Enough about forerunners, let us begin the count with Euclid.

He lived and worked in a time that is curious in the extreme.

In the year 323 BC, as a result of an acute fever or of immoderate drinking, or simply due to a goodly portion of poison, the king of mortal kings, Alexander of Macedonia, though a relatively young man of thirty-three years yet worn and weary, departed for a meeting with his father Zeus.

The demigod was hastily disposed of, for affairs of state demanded attention. The empire had to be divided, and this was no ordinary empire. Within a matter of ten years, lands had been conquered that exceeded the tiny poverty-stricken Macedonia by hundreds of times.

How and why this came about is not of great interest to us here. There were many reasons. One of which, as it will be recalled, was that Alexander of Macedonia was a hero.[1]

Be that as it may, the world had changed in ten years. Its boundaries had expanded four times over, and now came the time to digest the spoils. One thing was clear, it was too much for one heir. And it would have been ridiculous to have given it all to the infant son of Alexander who was born several months after the father's death, or to the second heir, Alexander's imbecile step-brother. And so it came about that the empire was ripped to pieces by those beloved generals that Alexander had not had time to execute.

They concluded an eternal peace, pledged just as eternal a friendship, drank heavily rejoicing, clasped each other's hand in a firm masculine shake and went their ways to begin slaughtering and fighting among themselves.

The times were exciting. Kings grew up like mushrooms after a rain

---

[1] An allusion to *The Government Inspector* by Nikolai Gogol.—A.S.

and were wiped out just as quickly. The lawful heirs, with no more guilt than their origin, were by the beginning of the second decade either cut down or strangled. The dynastic wars and slaughter continued for a few more decades.

That was how the extremely interesting era of Hellenism got under way.

In this fracas, luck was on the side of the circumspect Ptolemy, who sliced off Egypt as his share.

He rather successfully interfered in the quarrels of the Diadochi (heirs), more or less reasonably held in rein (within the confines of Egypt) his audacious Macedonians whose sword-points kept him in power. He did not oppose the worship of black cats and crocodiles so dear to the hearts of the local scholars, and he himself became a god in accord with the position he held (after all, he was a pharaoh). He plundered the country, and did so efficiently. True, this could not really surprise the Egyptians. He encouraged trade, killed off—on a moderate scale—those dissatisfied with the way things were going, and pampered the bureaucratic apparatus... He got to like the banks of the Nile, especially this new town to which Alexander gave the original name of Alexandria.

His heirs gradually settled down and the dynasty took firm root. What is more, it gave to the world Cleopatra, and literature had an exciting topic for two thousand years.

The very first Ptolemy, called Ptolemy Soter, and all succeeding Ptolemies stand out as patrons of the sciences. It is hard to say today what motives lay behind this sudden interest that the Ptolemies took in the sciences.

Perhaps it was a kind of intellectual coquetry. It might be that in attracting mathematicians and philosophers, Ptolemy I was aping Alexander—after all, Alexander was a pupil of Aristotle and he valued men of learning (true, his liking took very peculiar forms). Then of course it may even be assumed that there was hope of putting the wisemen to some kind of practical use. This was rather doubtful though.

Let us put aside guesswork and note facts, only facts.

In the third and second centuries BC, Alexandria had become the principal centre of learning of the Hellenistic world. And the most magnificent institution of learning was the celebrated Museum of Alexandria with its famous library. Unfortunately, it was plundered many times, and to complete matters, all 70,000 scrolls perished in a fire started in the seventh century by some furious Arabian calif.

Incidentally, it seems that the calif was really not so much to blame. The first one to have a hand in it was the great Caesar—Gaius Julius Caesar, a fairly decent writer of prose and also and mainly a general and political demagogue with boundless ambition.

Too, there are extremely weighty reasons to believe that in the main the work was that of the early Christian church (it was extremely tolerant of other faiths already at that time), which got out ahead of the simpleton calif by about two hundred or so years. All the calif had to do was clear away the remains. However that may be, the very best work of the Ptolemies awaited an unpleasant fate.

At any rate, if we are to remember the Ptolemies for what good they did, it is for their patronage of learning.

Human history has known many kingdoms and more kings. It may be that historians will trace the relationships of the doings of one and another despot and subsequent events. But the living memory of the people carries along a negligible percentage of all this crown-bearing horde. And what memory there is, is most often bad.

Those that stand out most—their luck—are cut-throats and adventurists like Tamerlane or Napoleon.

But the role they play today in our life is practically nil.

Since I have delved rather deep in these ancient variations on the topic of the frailty of earthly kingdoms and their glory, let me conclude with a parable.

Some few decades prior to the invasion of Mexico by the Spanish, a certain Aztec leader with a totally unpronounceable name (let us call him X) united all the tribes into a kingdom, thus to some extent eliminating the feudal fragmentation of the land. It was naturally thought that the kingdom and his dynasty would last for long centuries. X himself ruled long and happily.

But Mexico was soon visited by the gangsters of Cortez, and all that was left of the Aztec empire was the ruins of what were once magnificent cities. But that is only half the story.

King X (the cacique, to be precise) quite naturally had a harem, for King X adored the female sex.

He was indeed an extraordinary man, a talented lyrical poet. Most naturally, he wrote poetry for his numerous wives in between attending to the affairs of state. It is his songs that can still—today—be heard in the villages of Mexico. We may rejoice once again that genuine works of art are always more lasting than any empire.

It is probably worth recalling the name of the poet, but alas I only remember that it is very long and hard to pronounce.

So Ptolemy the First, Soter, invited Euclid to Alexandria. There Euclid wrote the *Elements*, a unique book unparalleled in human history.

Again I have to admit that practically nothing definite is known about Euclid the man. Of course, a couple of apocrypha we have.

It is said that at first Ptolemy himself wanted to master the intricacies of geometry. But he soon found that the study of mathematics was too onerous a burden for a pharaoh. Then he invited Euclid and asked him (oh, surely, as one gentleman would another): "Is there not some easier way of grasping all the secrets of learning?" To which Euclid, the story goes, replied proudly and not so politely: "There is no royal road to geometry." We do not know whether Ptolemy continued studying geometry. Most likely he found comfort in the business more suitable to kings (receptions, hunting, drinking and his harem).

The other story is that Euclid was approached by one young pragmatist, who asked: "What is the practical use of studying the *Elements*?" Whereupon Euclid, touched to the quick, called a slave and said: "Give him threepence, since he must make gain of what he learns".

True, both stories reflect the traditional view the Greeks took of wisemen and mathematics so well that one is not inclined to believe either of them.

The first story may be very pleasant to the modern ear, but the second is rather objectionable. Taking into consideration that the Greeks thought... well, on the other hand, I guess I just don't know what the Greeks really thought. Some, I'm afraid, thought one way, and others another. We know (that is, we think we know) that they despised all practical application of mathematics. And it would indeed seem that the philosophical works of those ages (particularly among the followers of Plato) corroborate as much. Repeatedly, in fact. That may be. And it may be true. But the greatest genius of mathematics of antiquity—Archimedes—was also a physicist, an experimentator rather than a theoretician. There is more. He was also a first-class military engineer who spent many years and much energy in building an impregnable fortress out of his home town of Syracuse.

Of course, Plutarch, taking upon himself to justify Archimedes, explained shamefacedly that all these things were toys, intellectual baubles of the philosopher. One however does not need to be perspicacious to realize that to plan the system of defence of your town equipped with weapons of your own invention is more than simple recreation. I repeat, an absolutely—for those times—impregnable system of defence.

Just one little aside: Archimedes and his work is a beautiful instance demonstrating that in those distant naive times physics and other sciences played just as important a role in affairs of war as they do today.

As to the actual attitude of the Hellenistic world to the practical utilization of mathematical knowledge, we are not sure.

Generally speaking, sweeping statements about that long-past epoch are always a bit irritating. We know so little; far too fragmentary and accidental are the facts of that past for us to speak definitely about the psyche and the customs of those people. I fear I am walking on thin ice myself taking up a discussion of matters with which I am not so very familiar. But before returning to geometry and Euclid I will permit myself just one remark, it is so tempting.

There seem to be two extreme trends in appraisals of the ancients.

Either the Greeks (the Hellenistic world, in particular) are idealized, and the protagonists of this view lament bitterly the decline of morals over the past 2,500 years and the forever-lost days of the childhood of mankind when people were pure, naive and devoid of guile (this is very popular among sophisticated intellectuals with a humanitarian slant).

Or—to take the other extreme—one needs only switch on the vacuum cleaner or television set to realize modern man's total moral supremacy over representatives of any earlier civilization. That often is the reasoning of technologists, the military and other exact professions (you will excuse me for not including physicists in this group).

As is so often the case, the disciples of the opposing camps are, essentially, at one in their lack of any desire to investigate the matter seriously, and they rely almost completely on haphazardly amassed impressions.

There is in addition a sceptical school of thought whose adherents claim that human beings have been the same over the ages and that man's intellect and moral qualities have not changed substantially during this measly period of only 2,500 years.

The author sides more with this latter view, though judging by what he—the author—has read, humankind has, over the past 2,500 years, been improving slowly but surely. One would like the advance to be somewhat more active. But that is a different question.

It is now probably time to explain to the tired reader why a book devoted to geometry digresses time and again into discussions about everything but geometry.

I will do that and then we will return to Euclid.

This is in lieu of an introduction.

Yes, what follows is going to be about non-Euclidean geometry and about the general theory of relativity, the creation of which may, without stretching the point very far, he considered the logical culmination of the whole story of the fifth postulate.

But what strikes me as most interesting in this story, is not geometry or the relativity theory. Ultimately, the entire epic about the fifth postulate is just as much witness to the power of human thought as it is to the remarkable, almost fantastic narrow-mindedness of mathematicians. No wonder, incidentally, that Max Planck permitted himself the perhaps overly categorical but, generally, correct statement that "in comparison with the theory of relativity, the construction of non-Euclidean geometry is no more than child's play". Let us, however, not be too jubilant. The important thing is something else.

The most important thing, the most instructive thing, and if you like, the most touching thing is that this story, which we now begin, is in fact a symbolic illustration of one of the best qualities that mark off human beings from the other primates and unite all races into a single species. The reader has guessed what the author is about: he sings the praises of the endeavour to find out what the world is like in which we live, how our universe is constructed. And he finds that the internationalism of earthlings, the internationalism of epochs, countries and peoples will eternally stand against the just as eternal coalition of narrow-mindedness, the

brotherhood of despots, go-getters, conquerors, climbers, grabbers, and the worst portion of sports fans.

If one could imagine for a moment the fantastic picture of Euclid, Omar Khayyam, Gauss, Lobachevsky and Einstein all in one room together, it is hardly likely that any of them would feel the need to gossip about common acquaintances or, for lack of a topic of conversation, to say, "how about a couple of jokes."

But on the other hand, one has to admit, albeit grudgingly, that the jokes of Euclid's time (with slight modifications for local colour, of course) almost fully exhaust the spiritual arsenal of very many of our contemporaries.

Incidentally, it is not worth idealizing either learning or the priests of learning. Hundreds upon hundreds of brilliant minds have turned out to be quite amoral personages.

And perhaps one of the most attractive features of this whole story is that just as non-Euclidean geometry logically culminated in the general theory of relativity, so the galaxy of mathematicians who worked on it—as a rule, not only remarkably talented but humanly interesting people—ends with Einstein.

But let us return to Euclid!

To begin with, a few words—the stronger, the better—about all the beasts that liquidated the Alexandrian library. If it had not been destroyed, we would now know scores of times more about the Greek and Roman worlds than we do.

We would probably know about Euclid as well. But, unhappily, as of today practically the most fundamental source on Euclid is Proclus Diadochus of Constantinople, a geometer who wrote an exceedingly detailed *Commentary* on the first book of the *Elements*. Since we are referring to sources, a slight remark will not be amiss.

When we turn to the history of antiquity, the effect is somewhat like that of regarding a chain of mountains from an aeroplane. Everything is smoothed over, distances contract, and small features vanish. Only the general overall picture remains.

Involuntarily we look upon all Greek mathematicians as almost contemporaries. Note, then, that Proclus (412–485 AD) lived seven hundred years after Euclid—a span of time much greater than that which separates us from Ivan the Terrible. Quite obvious then that the facts at Proclus' disposal concerning the life of Euclid were fragmentary and haphazard.

There is another author who lived a few decades before Proclus. He was the Alexandrian mathematician Pappus. He wrote of Euclid describing him as mild, modest and, at the same time, independent. Both relate the incident with Ptolemy. "Exact" biographical data are mostly based on the remarks of an unknown Arabian mathematician of the twelfth century: "Euclid, son of Naucratus, the son of Zenarchus, known by the name of Geometer, a scholar of olden times, of Greek origin, lived in Syria, born in Tyra...."

That is all.

The man dissolved in the ages without a trace. What remains is his work.

We repeat, the *Elements* is a book without parallel. For over two thousand years it was the principal and practically sole manual on geometry for scholars of both the Occident and the Orient. As late as the end of the 19th century, many English schools taught geometry on the basis of an adapted edition of Euclid's *Elements*. There can hardly be a more eloquent witness to its popularity. In this sense, only the Bible can compete with the *Elements of Geometry*. But unlike the Bible, the *Elements* are a rigorous system of logic. To be more precise, Euclid ever strived towards such a system.

We can presume that Euclid was a follower of Plato and Aristotle. Plato, as you recall, demanded a strictly deductive construction of mathematics. At the foundation were axioms: the basic propositions that were accepted without proof; from then on, everything had to follow with utmost rigour from these axioms.

That was the ideal that Euclid attempted to accomplish. Attempted, because from the viewpoint of today literally his whole axiomatics is unsatisfactory.

But that is easy to say now, after 25 centuries of investigations. In its day, Euclid's logic left an overwhelming impression.

Attempts had been made before Euclid to describe geometry on the basis of an axiomatic method. Not bad attempts, either. But we can assuredly say that Euclid's work was the most successful, as witness the unprecedented popularity of his book already in ancient times—a popularity that brought the book down through the ages to us.

One can say all kinds of harsh (and true) things about Euclid's axiomatics. But one should never forget that the scheme itself became, since that time, the canonical model for constructing every branch of mathematics. And of course one must never forget that the *Elements* present an excellent piece of writing by a skilled master, a perspicacious scholar and a magnificent teacher. That explains and justifies the universal admiration of mathematicians for Euclid and his *Elements*. Let us add that this book brought to the field of mathematics scores of young men who later became the world's greatest mathematicians.

The influence of Euclid has been amazing throughout the ages and throughout the world. Take one of the most prominent mathematicians of the Renaissance, Cardano, who, it must be added, was a rabid adventurist (not to say scoundrel) but there is no getting around his mathematical talent and culture. Here is how he admired Euclid's *Elements*.

"The irrefutable strength of their dogmas and their perfection are so absolute that not a single work can justifiably be compared with them. As a consequence, there is such a light of truth reflected in them that, apparently, only he is capable of distinguishing the true from the false in the intricate problems of geometry who has mastered Euclid."

In the middle of the 19th century a distinguished British geometer had this to say: "There has never been a system of geometry, which, in its essentials, has differed from the plan of Euclid; and until I see such with my own eyes, I will not believe that such a system can exist."

True, it must be said that in the middle of the 19th century, that geometer could have reasoned more progressively and these words, aside from worship of Euclid, demonstrate the author's own hidebound conservatism.

We could go on citing numerous other writings in the same vein, but we will confine ourselves to what is probably the most brilliant demonstration of the effect the *Elements* had on literally all fields of thought. Benedict Spinoza, celebrated philosopher of the Western world, borrowed the entire plan of his basic work, *Ethics*, from Euclid.

Perhaps the authority of Spinoza is not convincing enough to some readers. If it isn't, let me mention Isaac Newton.

His fundamental work, the *Principia* (The Mathematical Principles of Natural Philosophy) copies Euclid both in title and outline. Axioms form the starting point from which everything else follows. The similarity may be continued because Newton's axiomatics turned out to be just as ephemeral as did Euclid's.

One final piece of information. By the year 1880, the *Elements* had appeared in 460 editions.

Perhaps a word is in order, at this point, about the axiomatic method itself.

It was only at the beginning of this twentieth century that we achieved a perfectly clear and rigorous understanding of deductive schemes. In the main, merit for this goes to the great German mathematician Hilbert.

In a rough and greatly simplified form, the matter stands as follows. We confine ourselves in what follows to the concrete material of geometry so as to avoid too many abstractions.

1. A List of the Basic Concepts

The foundation of the geometry are Basic Concepts (alias, *Primitive Notions*). These are the result of a prolonged experimental study of nature, a study both intricate and confused and nebulous and more. But Basic

Concepts carry no trace of that. They represent a certain abstract reflection of reality.

Basic Concepts have no definition. They come ready-made, as you might say. This is natural enough. To define them, one would need some other notions, which in turn can only be defined with the aid of... and so on *ad infinitum*. One has to start somewhere. As the French say, "in order to make a rabbit stew, one at least has to find a cat".

So we have the Basic Concepts. Mathematicians have a delightful way of putting it: these are elementary entities that are not defined, they are simply stated. One can add here that, in the modern axiomatics of geometry, the Basic Concepts are divided into two groups:

(a) Basic Terms;

(b) Basic Relations.

One must say that today there are at least two essentially different axiomatic schemes. In what follows we will use the scheme in which the Basic Terms are as follows:

(1) point, (2) straight line, (3) plane.

Now let us see what the Basic Relations are. They are formulated as:

(i) to contain, (2) to lie between, (3) motion.

Now that the Basic Concepts have been established, we can go over to the second stage.

2. Basic Axioms

For our Basic Concepts we make a set of assertions that are accepted without any proof. These are axioms. Speaking in strictly formal fashion, it is only the axioms that fill our Basic Concepts with live content. Only they impart life. Without the axioms, the Basic Concepts are devoid of any content. They are nothing. Amorphous ghosts. The axioms define the rules of the game for these "ghosts". They outline a logical order. The mathematician can say only one thing about his Basic Concepts, that they obey such and such axioms. That and nothing else! And all because the mathematician does not in fact need to know what he is talking about. He demands only one thing: that his axioms be satisfied.

That and nothing else!

Once the axiomatic method has been elaborated to perfection, geometry, speaking formally, is converted into an abstract game of logic.

The notions of point, straight line, plane, motion can mean anything, any entities.

Let us construct a geometry for them. We will then call our geometry Euclidean geometry if the axioms established for the "real" geometry of Euclid are fulfilled.

For example, *one and only one straight line can be drawn through two distinct points.* This is an axiom formulated in ordinary language.

If we were to adhere strictly to the terminology just introduced, we would have to make the statement:

There exists a single straight line containing two different points.

And so on in the same spirit. On the basis of this axiom, a good exercise is to prove the theorem: *Two straight lines may have not more than one point in common.*

At the present time, five groups of axioms are distinguished in Euclidean geometry. They are:

(1) axioms of connection;
(2) axioms of order;
(3) axioms of motion;
(4) the axiom of continuity;
(5) the axiom of parallel lines.

We prefer not to list all these axioms at the present stage, we will put them in the appendix to the next chapter, for, as Herodotus once said, nothing gives such weight and dignity to a book as appendices.

We shall have occasion to return to the axioms a number of times.

Meanwhile, we take up Stage 3.

3. The Basic Definitions

With the aid of the Basic Concepts we construct more complicated ones. For instance, *an angle is a figure formed by two half-lines (rays) containing a single common point.*

A careful reading of this phrase will make it clear at once that one complex concept (namely, ray, or half-line) is used in the definition of an angle.

Obviously, we should have given the definition of this notion earlier with the help of the Basic Concepts. This is rather easy to do. The reader can check to see how much he is now imbued with the spirit of deduction, and, armed with a list of axioms, can try to solve the problem.

If it turned out that in employing the Basic Concepts, it was impossible to define a ray, then one would have to place this notion in the category of Basic Terms.

In general, all other notions are defined with the aid of the Basic Concepts, and also (take note!) of the axioms established for them.

There remains the last stage:

4. Statement of Theorems. Proof of Theorems

We formulate certain propositions (theorems) for our notions (basic and non-basic) and prove them. That, properly speaking, forms the subject matter of geometry.

I should like to repeat once again that, when stated in those terms, geometry is converted into an absolutely abstract game, like, say, chess.

There, too, we have Basic Concepts, called chessmen. The axioms are the collection of rules of the game. Finally, there are theorems. Actually, only one theorem: how to checkmate the opponent.

In solving this "theorem", a player proves dozens of lemmas (auxiliary theorems) in the course of a game, each time selecting the best (in his opinion) move in a given position.

True, there is a difference between games and geometry. It consists in the fact that the partners very often produce incorrect proofs. In chess, for example, no strict logical criteria for evaluating every move or position have yet been evolved. In geometry they have. Here, it is in many cases possible to establish whether a newly formulated theorem contradicts earlier

theorems, and hence runs counter to still earlier ones, and consequently...
Unravelling the roll to the end, we arrive at two possibilities: either we
have erred in our reasoning, or the theorem just formulated is erroneous.

The former possibility is of little interest to science: the only thing it
shows is that we have handled the mathematics poorly. But in the latter
case there is often a definite and very important result. If we have become
convinced that our hypothesis (theorem) is wrong, then other theorems are
right, namely those which contradict our own. If there is only one such
contradicting theorem, then we have proven it by our reasoning.

This last paragraph, though perhaps rather nebulous and abstract in
form, is an explanation of a scheme that is very common in geometry (and
mathematics generally). It goes by the name of *reductio ad absurdum*, or
indirect proof.

Coming down to earth again, let us take a specific case to prove.

Let there be two perpendiculars dropped onto a straight line. Using
radian measure of angles and writing $\pi/2$ in place of 90 degrees, we find
two, and only two, variants: either the perpendiculars meet at some point
C or they do not intersect at all. Let us prove that the second theorem is
correct. We do so by the method of *reductio ad absurdum*.

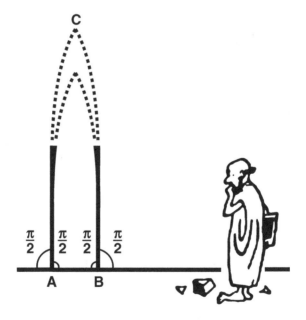

Assume that the first supposition is fulfilled and that the two perpendicular lines intersect. Then we have a triangle ABC (for triangle we will use the symbol $\triangle$, for angle the symbol $\angle$). The remarkable thing here is that the exterior $\angle B$ is equal to the interior $\angle A$. And of course the exterior $\angle A$ is equal to the interior $\angle B$.

But there exists a theorem (at the moment, we will assume it to be true): *An exterior angle of a triangle is always greater than any interior angle not adjacent to it.*

Our triangle does not satisfy this theorem. Hence there can be no such triangle. Consequently, we are in error.

A check of the reasoning shows that everything is correct. Hence, the error was made at the very beginning, when it was assumed the perpendicular lines intersect.

Thus, perpendicular lines do not intersect. That has been proved rigorously. Euclid called nonintersecting lines parallel lines. For the time being we too will use this terminology.

To summarize, then, we have found that two straight lines perpendicular to a common straight line are parallel. We should also prove that these straight lines do not intersect in the lower half-plane either. But that would simply be repeating the preceding proof, and our time is limited.

In carrying out the proof we relied on the theorem of the exterior angle of a triangle. The alert reader will of course see that the whole example is very important for what is to follow, and so without any more digressions we will prove this theorem too. It is of ultimate importance to us, and to the entire story involving the fifth postulate.

Appreciate the suspense: the postulate itself has not yet been formulated in any way.

Still we must say: the whole story of the fifth postulate started with this very theorem.

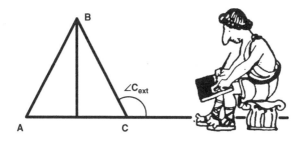

Let there be a $\triangle ABC$. Look! The exterior angle C_{ext} is clearly indicated

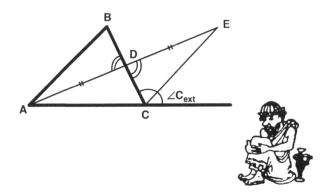

by the arc. We shall prove that it is greater than any interior angle not adjacent to it; that is to say, greater than $\angle A$ and greater than $\angle B$. We start with $\angle B$.

Divide the side BC by the point D into two equal parts and draw a straight line through A and D.

On this line, mark off a segment DE equal to AD and connect points E and C by a straight line.

The triangles ABD and DEC are equal (*congruent*, a mathematician would say). Indeed, we have by construction $AD = DE$ and $BD = DC$. The angles $\angle CDE$ and $\angle ADB$ are equal because they are vertical angles.

Hence, the triangles are equal on the basis of a familiar criterion.

But then $\angle B$ (or $\angle ABC$) is equal to $\angle BCE$! And note! The angle $\angle BCE$ is only a part of the angle C_{ext}.

Thus, the entire angle C_{ext} is greater (naturally greater, for the whole is always greater than one of its parts) than the angle B.

Some doubt remains about $\angle A$. It is immediately felt that our construction will not be of any particular help, since in the figure $\angle A$ is cut into two parts. It would be good to put it in the position of $\angle B$. Perhaps we should draw a straight line from the vertex B and repeat our construction and proof. But then $\angle C_{\text{ext}}$ will be placed differently.

A complete analogy would result if we prolonged the side BC and regarded the new external angle N.

$\angle N$ is of course greater than $\angle A$. We have just proved that.

An inspiration! The angle N equals the angle C_{ext} because they are vertical angles.

That is all.

An exterior angle of a triangle is greater than any interior angle not

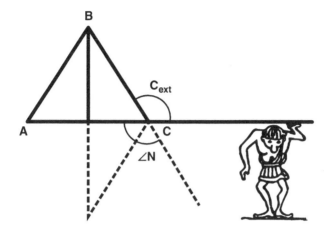

adjacent to it. We have proven this and we can now cross out the doubts we had on page 27 about the validity of the theorem.

If we go over the path traversed with exceeding care... And if we check to see which axioms have been utilized in the proof of the theorem of the exterior angle... To do this we would of course have to verify the axioms that were used in proving the theorems of the congruence of the triangles and the equality of vertical angles.

Now if all that were done, we would find that we have utilized practically all of the axioms.

But nowhere have we taken advantage either of the very notion of non-intersecting (parallel) straight lines, nor (all the more so!) of theorems or axioms concerning such straight lines.

The reader can easily verify this by taking the list of axioms and analyzing all the Concepts that are needed for the theorem of an exterior angle and for all auxiliary theorems.

Our detour has been too long and it is time to return to the axioms.

First, let us figure out what logical requirements they must satisfy.

Only two: (i) completeness and (2) independence.

The first signifies that there must be a sufficient number of axioms to prove or disprove any possible assertion concerning our primary Basic Concepts or the more complex Concepts built up from them.

The second implies that we did not take too many axioms. We have just exactly the number we need. And not a single one of the axioms can be proved or disproved with the aid of the others.

Both these demands may be formulated in a single statement. The axioms must be necessary and sufficient.

Necessity is a requirement of independence.

Sufficiency is a requirement of completeness.

To put it very roughly, the requirements of necessity and sufficiency signify that there must be exactly the number of axioms as is needed, neither more nor less.

Now for one very important refinement.

From the independence of the axioms there follows straightway their consistency. Indeed, if in our development of geometry we at some stage arrive at a theorem that contradicts the rest, this will be a clear unpleasant indication that there is something wrong in the foundation. Namely, that one axiom (or several) contradicts the rest. And if there is an inconsistency, that means they are not independent.

Actually, all these logical arguments are extremely simple. But in a first reading they may appear rather involved. My suggestion is for the reader to go over them once more.

For the present I would emphasize once again that the requirement of independence of axioms is stronger and more rigid than the requirement of consistency.

The axioms may be consistent, but from this consistency it does not yet clearly follow that one of them might not be a corollary of the others. Perhaps it is a theorem. Naturally, when a mathematician proposes a system of geometrical axioms, he is obliged to prove their independence! Let us stop our chain of reasoning at this point. There will be time and opportunity to return to them again. We will not miss the opportunity and will not lose time either, of that I am certain.

Although everything that has just been written is rather simple, and I am positive the reader thinks so too, Euclid did not know any of it. Intuitively he felt it all, but he could not formulate it in a clear-cut logical scheme.

Now a rigorous statement of the problem of the independence of axioms or the rigorous introduction of the Basic Concepts was generally beyond the ken not only of the Greeks but of mathematicians in all ages and peoples right up to the 19th century.

Both the axiomatics and the proofs provided by Euclid are actually a rather varicoloured mixture of intuition and logical lacunae—if one regards them from the standpoint of today.

Yet, on the other hand, Euclid advanced so far and so crucially along the road to rigorous logic that all other textbooks, and all other "elements" abundant in antiquity paled completely when compared with the *Elements*.

When the Greeks spoke of Homer they simply said the "poet", and when the Greeks recalled Euclid they said the "maker of the *Elements*".

All predecessors on the deductive pathway of geometric constructions were forgotten.

There remained the *Elements* and their creator Euclid.

Although the thirteen books written by Euclid are believed to contain mainly the results of others, and for this reason the question is often debated as to whether he may be classed as one of the greatest mathematicians, he was without doubt a teacher of the first magnitude. We may also add that he was apparently an inspired and versatile scholar, for in addition to the *Elements* he also wrote *Elements of Music, Optics, Catoptrica* (theory of mirrors), *Data Phaenomena* (a work on astronomy), *Introductio harmonica*; then also works that did not come down to us and disappeared: the *Porisms* (in three books), *Conics* (in four books), *Perspective* (in two books), *Surface Loci, On Divisions of Figures* and a *Book of Fallacies*.

A very impressive list.

Most of the books, it is true, make no original contributions, but the output of work is tremendous. Incidentally, the Data was highly valued by Newton, which is a rather solid recommendation. Euclid apparently advanced substantially the highly complex and exciting division of Greek geometry devoted to the teaching of conic sections. However, he did not include these results in the *Elements*, since there was a current view that this branch was unworthy of "pure mathematics, whose aim is to bring man closer to God".

It was again Plato who decided why precisely the theory of conic sections did not bring one closer to the divine. The point was that Plato viewed as heresy the use in geometry of any instruments other than the compass and the straightedge, or—what is the same thing—the use of loci other than the circle and the straight line (and such loci were needed in the study of conic sections). Plato passionately denounced the brilliant geometrician Menaechmus (incidentally his friend), who demonstrated that the solution of the notorious problem of duplicating the cube, also that of trisecting an angle, is found rather simply if use is made of new geometrical instruments.

Plato maintained that all of that "spoils and destroys the good of geometry, for geometry thus strays away from incorporeal and mentally perceivable things and moves towards the sensorial, making use of bodies that require the application of instruments of vulgar handicrafts.

Obviously this rebuke frightened poor Euclid, and his work on conic sections vanished without a trace.

There would seem to be something in the *Elements* dealing with regular solids (polyhedrons) that belongs to him. In the thirteenth book it is proved that there exist only five different types of such solids. This is a brilliant, unexpected, celebrated... classical result.

Generally speaking, there is much in the *Elements* other than geometry. They contain certain essentials of the theory of numbers and the geometrical theory of irrational quantities. The three last books are devoted to solid geometry. Every division is preceded by axioms and postulates.

Properly speaking, plane geometry is explained in the first six books, the very first of them beginning with axioms and postulates.

Mathematical historians are still not in agreement as to how Euclid distinguished between axioms and postulates.

Generally, to Euclid, axioms (which he calls "general attributes of our mind") are truths that refer to any entities (not only geometrical). For example, if A is equal to C and B is equal to C, then A is equal to B. Here, A and B may be numbers, segments of lines, weights of bodies, triangles, etc.

Postulates, on the other hand, are purely geometrical axioms. For instance, Euclid's first postulate: "A unique straight line can be drawn from any point to any other point."

Euclid also has Basic Concepts (primitive notions).

There is hardly any reason to give his entire system of axioms—we have said that a dozen times—because it is quite unsatisfactory. There are, properly speaking, six axioms in Euclid's plane geometry, and we shall not mention them. But the postulates are worth noting. Here are the first four.

It must be required:

I. That a straight line may be drawn between any two points.

II. That any segment of a straight line may be extended indefinitely.

III. That about any point as centre a circle with any radius may be drawn.

IV. That all right angles be equal.

For the time being we shall not stress what is bad in these postulates. As Nikolai Lobachevsky once said, forgive Euclid and the *Elements* all their "primeval shortcomings". The important thing for us at present is that all four postulates are very elementary in content. Here Euclid postulated absolutely natural, comprehensible truths that are part and parcel of our consciousness and our intuition. All is well and good, and... then we come to the fifth postulate.

Chapter 3

The Fifth Postulate

The fifth postulate reads:

If two lines are cut by a transversal and the sum of the interior angles on one side of the transversal is less than two straight angles (2d, or 180°), the two lines will meet if extended and will meet on that side of the transversal.

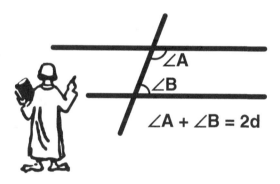

That's a formulation for you! First of all, what a lot of words. Secondly, what a lot of geometrical concepts. A person poorly familiar with the fundamentals of geometry will find it hard to understand anything. The postulate differs radically from all the others. It sounds more like a theorem. And not a simple one either. There is quite obviously something strange here. Before we go any farther, allow me to bow to Euclid.

Though I myself have, naturally, no proof, I am convinced that the fifth postulate was purposely formulated in this ugly form. Therein lies the great wisdom of the creator of the *Elements*.

Of all possible statements of the fifth postulate, Euclid chose the most intricate and cumbersome one. Why? To answer, let us see how he constructs geometry.

After the axioms and postulates, Euclid naturally proves theorems. He proves 28 theorems straight off without once using the fifth postulate. It is not needed. All 28 are indifferent to the fifth postulate, for, as they say, they refer to absolute geometry.

Among the twenty-eight there is also a theorem of the exterior angle of a triangle. In Euclid's list it is No. 16. The list terminates with, as you can easily imagine, No. 27 and No. 28. These theorems contain the so-called "direct theory" of parallel lines. We shall prove them together.

Let two straight lines he intersected by a third at points P and P_1.

It is asserted that *if the angle A equals the angle A_1, the straight lines are parallel.*

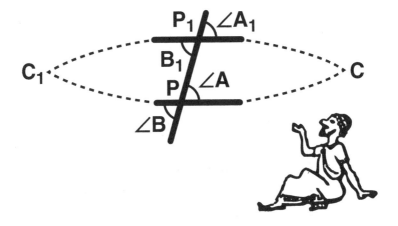

Working by the *reductio ad absurdum* method, we first assume that the straight lines intersect at point C. Then we get $\triangle PP_1C$, whose exterior angle A_1 is equal to the interior angle A not adjacent to it. But this is impossible. The theorem proven in the preceding chapter that *an exterior angle of a triangle is always greater than any interior angle not adjacent to it* does not allow this to occur! Hence, the straight lines cannot intersect when extended to the right.

There is a second possibility. The straight lines intersect at point C_1. Then we get $\triangle PP_1C_1$ for which $\angle B$ is an exterior angle and $\angle B_1$ is an interior angle not adjacent to $\angle B$.

But $\angle B = \angle A$ and $\angle B_1 = \angle A_1$, they being vertical angles. On the other hand, by hypothesis, $\angle A = \angle A_1$. It follows that $\angle B = \angle B_1$.

We are almost there.

For the hypothetical triangle PP_1C_1, the angle B is an exterior angle and the angle B_1 is an interior one not adjacent to it. And they are equal. Which is impossible. Consequently, $\triangle PP_1C_1$ cannot exist. Hence, the straight lines do not intersect in point C_1 either.

That completes the proof of the theorem.

It is obvious to the reader that $\angle B$ and $\angle B_1$ were introduced so that for the hypothetical triangle PP_1C_1 we could completely duplicate the reasoning that we carried on for $\triangle PP_1C$ (the first triangle).

Now, so as to completely repeat Euclid, let us introduce four more angles into our drawing. A glance at the figure will indicate which ones.

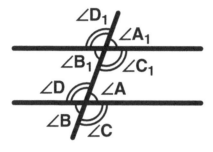

From the equality $\angle A = \angle A_1$ there straightway follows a whole family of equalities.

(1) $\angle B = \angle A_1$; $\angle C = \angle D_1$ these angles are called "opposite exterior angles".

(2) $\angle A = \angle B_1$; $\angle D = \angle C_1$; these are called "opposite interior angles".

(3) $\angle D = \angle D_1$; $\angle C = \angle C_1$; $\angle B = \angle B_1$; and, naturally, $\angle A = \angle A_1$. All these angles are called corresponding angles.

(4)

$$\angle D + \angle B_1 = \pi \,,$$
$$\angle A + \angle C_1 = \pi \,,$$
$$\angle C + \angle A_1 = \pi \,,$$
$$\angle B + \angle D_1 = \pi \,.$$

Here, we have the sums of interior and exterior angles *on one side*.

Obeying the generally accepted order of things, I listed all twelve equalities and now regret it. So many can easily obscure a clear matter. Any single one would suffice. The other eleven follow from that! We started with the equality $\angle A = \angle A_1$. But any other one would have been perfectly suitable.

We proved that if any one of the twelve equalities is fulfilled, then the straight lines are parallel. This is the essence of Euclid's two theorems, No. 27 and No. 28.

Incidentally, it is worth recalling at this point that the theorem about the parallel nature of two lines perpendicular to a common straight line—the first theorem proved in this book—is a special case of the theorem just proved above.

Upon proving a theorem, the geometer always investigates the converse. In the converse, one proceeds from that which is proved in the direct theorem, and, naturally, the attempt is made to prove what is already given in the direct theorem.

One of the most common logical mistakes of beginners is connected with direct and converse theorems. It is casually thought by many that the converse of a theorem follows directly from the theorem itself.

To disprove this, let me cite the familiar reasoning of Captain Wrungel which I have kept in my memory all these years for just such a case.[1]

Direct theorem	**Converse theorem**
	(The theorem of Captain Wrungel)
Any herring is a fish	Any fish is a herring

In keeping with certain traditions of popular science literature, one adds at this point that the above example is just a joke. But I won't bother to do that.

Examples taken from geometry (Euclidean):

---

[1] From a popular Russian children's book.—A.S.

Direct theorems

I. If in the triangles ABC and $A_1B_1C_1$ the relations $AB = A_1B_1$ and $AC = A_1C_1$ hold and also $\angle A = \angle A_1$, then $\triangle ABC = \triangle A_1B_1C_1$.

II. Two lines perpendicular to a common straight line are parallel.

III. If $\triangle ABC$ is similar to $\triangle A_1B_1C_1$, then
$$\frac{AB}{A_1B_1} = \frac{AC}{A_1C_1}$$

Converse theorems

I. If $\triangle ABC = \triangle A_1B_1C_1$, then $AB = A_1B_1$ and $AC = A_1C_1$ and also $\angle A = \angle A_1$.

II. If two parallel straight lines are cut by a transversal, they are perpendicular to it.

III. If the proportion
$$\frac{AB}{A_1B_1} = \frac{AC}{A_1C_1}$$
holds for the triangles ABC and $A_1B_1C_1$, then the triangles are similar.

In Example IV, we combine four different theorems into one.

IV. If $\triangle ABC$ is an isosceles triangle $(AB = BC)$, then *(i)* $\angle A = \angle C$ and *(ii)* the altitudes, the medians and the bisectors of the angles A and C are equal.

IV. If in $\triangle ABC$ *(i)* $\angle A = \angle C$ or *(ii)* the altitudes or the medians or the bisectors of the angles A and C are equal, then the triangle ABC is an isosceles triangle $(AB = BC)$.

In these examples, all the direct theorems are correct. It is left to the reader to figure out whether the converse theorems are also valid. It is a curious fact, incidentally, that very often, though the converse is quite correct, it is far more complicated to find its proof than the proof of the direct theorem. Naturally, there is such a case in our examples as well.

One of the direct theorems in Example IV—the equality of bisectors in an isosceles triangle—has a simple proof, whereas the converse (which is an absolutely correct theorem) is somewhat of a tricky geometrical problem.

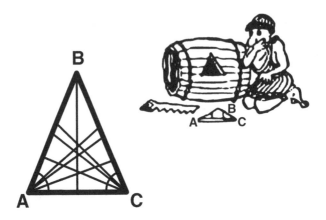

With the theorem of parallels proved, let us try the converse. We formulate it as follows:

Direct theorem of parallels	**Converse theorem of parallels**
If two lines are cut by a third and the equality $\angle A + \angle C_1 = \pi$ (or any other of 12 equalities given above) is fulfilled, then the lines are parallel.	*If two lines cut by a third are parallel, then the equality $\angle A + \angle C_1 = \pi$ (and any other of 12 equalities given above) is fulfilled.*

The converse theorem of parallel lines was taken by Euclid as the Fifth Postulate, though Euclid's formulation of the fifth postulate is somewhat different.

Recall the definition given at the start of this chapter. It is well worth the trouble. Here it is.

Postulate V. *If two lines are cut by a transversal and the sum of the interior angles on one side of the transversal is less than two straight angles (that is, the sum $\angle A + \angle C_1$ is less than $\pi = 2d = 180°$), the two lines will meet if extended and will meet on that side of the transversal.*

Both the purposefully cumbersome way in which Euclid introduced the fifth postulate and the fundamental 28 theorems which preceded it and which were proved quite independently of it, all go to demonstrate the amazing intuition of Euclid or of the one be borrowed the idea from (if that person existed).

I shall try to explain myself and substantiate my claim. This is all the more pleasant a task, since it will be quite impossible to refute what I have

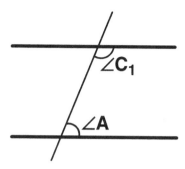

to say. There are no facts at all, thus opening wide all opportunities for an historico-psychological investigation.

Let us examine the initial data.

By the time the *Elements* were written, geometry had already grown into a mature, well-elaborated science.

Behind it lay three hundred years of development and dozens of intricate problems solved, and several tough unresolved ones like the duplication of the cube. Thanks to Plato and Aristotle, the deductive scheme was established, had gained recognition and was flourishing.

At this point, an historian of geometry can ennoble two score names of celebrated Greek mathematicians. Naturally, this refers to those scholars whose names have come down to us. For each one of them there are undoubtedly at least ten geometers of lesser magnitude whose names never reached us.

Practically all were in agreement that geometry should develop on the basis of axioms. Obviously, the majority were in full accord with Aristotle in that axioms and the basic notions should satisfy the requirement of being obvious. But, as Aristotle insists, the actual formulation of the axioms is a matter of too great a responsibility to entrust it to mathematicians. It is the supreme problem.

Naturally, then, only the most worthy were admitted to have a say about it.

Philosophers, in other words.

Whether the geometers believed Aristotle or not, is not the point; the point is that with Aristotle one agrees.

There can be no doubt that before Euclid's time attempts had been made (and numerous ones) to prove the *converse of the theorem of parallel lines*. And I personally think that by Euclid's time it was clear that two solutions existed:

1. To prove the converse theorem of parallels on the basis of the remaining postulates of geometry, and, by the rules of the game, without the introduction of any additional postulates.

The adherents of this school must have presumed that the converse theorem of parallels was nothing more than a complicated theorem that followed unavoidably from the other postulates.

2. To the four postulates it is possible to add a fifth such that the converse theorem of parallels would readily be obtained with its aid. And this additional postulate might be formulated in such manner that it would appear natural and obvious in the extreme.

It is hard to believe that the predecessors and contemporaries of Euclid—all brilliant geometricians of the age of flourishing learning—could not conjure up a whole galaxy of equivalent and "obvious" statements of the fifth postulate. It is hard to believe for the simple reason that some of them practically beg to be stated.

As far as the first path is concerned, no success was achieved either at that time or during the two thousand years following Euclid. Today, thanks to Lobachevsky, we know that success was out of the question. But that is what we know today.

All the more alluring was, most apparently, the second possibility: to propose an equivalent but simple and natural postulate—to smear over and cover up the unpleasant spot and calm down.

Numerous commentators of Euclid who dealt with the fifth postulate did just that—explicitly or in veiled form.

It is impossible to believe that such an outstanding mathematician as Euclid who profoundly researched the problem of the fifth postulate (and the entire construction of the first book of the *Elements* exhibits his particular attention to this problem), it is impossible, I insist, that he did not come across a number of equivalent and rather natural formulations of the fifth postulate. For instance, if we combine the direct theorem on parallels and the fifth postulate in Euclidean form, we immediately get:

A new formulation of the fifth postulate. *Through a point C lying outside a straight line AB, it is possible to draw in a plane ABC only one line that does not meet AB.*

This statement is usually attributed to the English mathematician Playfair (18th century), but, naturally, it was proposed by many commentators of Euclid many centuries before Playfair's time.

"Playfair's axiom" does look much more natural and attractive than

Euclid's postulate, doesn't it?

Here is another formulation. It is usually attributed to Legendre, though it too was employed earlier by European and Oriental geometers.

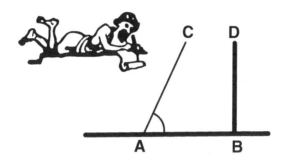

Legendre's postulate. *A line perpendicular to, and a line inclined to, a common transversal AB, located in the same plane, definitely meet.* (Naturally, from the side where the oblique line forms an acute internal angle with the transversal.)

Again a very pictorial assertion. In place of the Euclidean postulate we have a special case when one of the angles is right. It will readily be seen that this is quite sufficient to prove the postulate. Incidentally, for those who are making their first acquaintance with geometry, this is a worthy and rather involved problem that merits some attention. I will give a few hints and leave the rest to the reader.

Those who are not particularly excited about this proposition can simply skip the mathematics. But we will accept the Legendre postulate—a line perpendicular to, and a line inclined to a common transversal meet—and will prove the fifth postulate in the Euclidean form, which is the *converse of the theorem of parallels.*

First let us prove an auxiliary theorem, a Lemma (its statement will be given at the end of its proof).

Let two straight lines I and II be intersected by a third so that $\angle A < \pi/2$ and the sum $\angle A + \angle C_1 = \pi$. Then, by the direct theorem we know that these lines do not meet, for they are parallel.

Refresh in memory the proof of the direct theorem.

From point C drop a perpendicular onto the straight line I. This can always be done. The appropriate theorem was proved without a word about parallel lines.

Prove, given our condition $\angle A < \pi/2$, that the perpendicular CB is

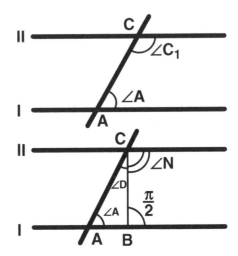

located as shown in the drawing.

Prove it by means of *reductio ad absurdum* and utilize the theorem on the exterior angle of a triangle.

We then have $\angle D + \angle N = \angle C_1$. $\angle N$ is not yet known.

Then we have $\angle A + \angle D + \angle N = \pi$. (Recall the hypothesis!)

Now consider $\triangle ABC$.

There are three possibilities for the sum of its angles:

$$\angle A + \angle D + \frac{\pi}{2} \;\; {\begin{matrix} > \\ = \\ < \end{matrix}} \;\; \pi.$$

Note: we cannot use the theorem that the sum of the angles of a triangle is equal to π. This theorem is a corollary to the parallel postulate.

First examine the hypothesis $\angle A + \angle D + \frac{\pi}{2} > \pi$. Compare this inequality with the equality $\angle A + \angle D + \angle N = \pi$ and obtain $\angle N < \pi/2$.

Now employing Legendre's postulate you get the straight lines I and II meeting on the right of point B. This contradicts the hypothesis. Consequently, the hypothesis is wrong.

Consider the hypothesis $\angle A + \angle D + \frac{\pi}{2} < \pi$. In exactly the same way show that in this case the lines I and II meet to the left of point B; then reject this hypothesis as well.

You have proven two important theorems at once:

1. *The sum of the angles of the triangle ABC is equal to π.*

2. *The angle N is equal to* 90°—this is our Lemma.

Now prove the converse parallel theorem by employing the following auxiliary construction.

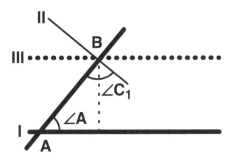

Given: when I and II are cut by a third line, let $\angle A + \angle C_1 < \pi$ and $\angle A < \pi/2$.

1. Drop a perpendicular onto the line I from point B.

2. Draw through B an Euclidean parallel line III, that is a straight line that satisfies the "direct theorem of parallels". Prove that it will pass as shown in the drawing.

3. Think for a moment and then again make use of Legendre's postulate to prove that the line II will intersect I.

You have thus proved Euclid's postulate. But do not forget that you made use of an equivalent postulate.

If you were somewhat embarrassed by the condition $\angle A < \pi/2$, convince yourself that it does not restrict the generality of your reasoning.

Now check through to be sure there are no errors in your reasoning.

The above proof has at least two noteworthy features.

First of all, we proved in passing that as soon as we took Legendre's postulate (the equivalent of Euclid's postulate) we found a triangle the sum of whose angles is equal to π.

Secondly, I have never read about this proof. I thought it up in a few minutes. Of course, I write this not because I am ambitious and hope to gain the admiration of the reader for my mathematical talent. My point was that cooking up such a proof was not a difficult matter.

The equivalence of the postulates of Legendre and Euclid can be proved still more simply and elegantly, in just two lines. All one needs to do is to derive first from Legendre's form of the fifth postulate its Playfair's version (through a given point only one line can be drawn parallel to a given straight line).

So, as you see, our theorem is unwieldy and unneeded. Its sole justification is that it suggests (not proves yet, but suggests!) another rather important theorem: *if the sum of the angles of a triangle is equal to π, then the fifth postulate is valid.* Besides, it is a useful exercise. Still more important—most important in my opinion—is the fact that such "investigations" demonstrate how the very first naive steps take us directly to ever new equivalents of the fifth postulate. And, of course, there can be no doubt that our simple chain of arguments was tried out by any number of commentators of Euclid.

But, being now convinced how easy it is to simplify the statement of the fifth postulate, we unwittingly ask: why did not Euclid do this himself.

I'm sorry but I cannot help myself. The situation demands a series of rhetorical questions, like:

Can it be that Euclid did not try to prove his theorem?

Is it possible that a scholar of that magnitude, such a perspicacious analyst could not obtain a few elementary corollaries and choose for the postulate the more natural and obvious one?

How can it be that he, a follower of Aristotle and Plato, let such an opportunity pass by?

How is it possible that he ruined the whole harmony of geometry thus bringing upon himself the ire of the immortal gods of Olympus?

Can it be that any one of a host of commentators was able to penetrate deeper and better into the problem than he?

The absurdity is so obvious... The most likely version is the following.

Euclid, like his predecessors, undoubtedly did attempt to elevate the fifth postulate to the rank of a theorem and prove it without involving any supplementary assumptions.

Taking into account the exceptional position of the fifth postulate in the *Elements* and also the notorious 28 theorems that preceded it, one can conclude with assurance that this problem worried Euclid and that he paid very special attention to it.

Recalling that all the methods of elementary geometry were fully elaborated in Euclid's day, recalling for instance that studies in the theory of conic sections were immeasurably more complicated than most of the reasoning involved in the fifth postulate, recalling (once again) that the fifth postulate—in the form that Euclid presented it—is a challenge to the demands of Plato and Aristotle, outright effrontery, recalling that Euclid, judging by everything we know, was their true follower..., and, finally, recalling that Euclid was a brilliant geometer... Recalling all this, we arrive at one and only one conclusion.

In the process of vain attempts to prove the fifth postulate, Euclid most likely found several equivalent formulations. Simple ones. Obvious ones. But Euclid knew where to stand.

On the one hand, he clearly understood that it would be impossible to prove the postulate without invoking some equivalent assumption. On the other hand, not one of the equivalent forms of the fifth postulate satisfied— to his liking—the requirement of being self-evident. And so he concluded that the situation was very sad and the problem remained unresolved. And, like an honest geometer, he decided to emphasize the fact that the fifth postulate was a rejected, despicable monster in the closely knit family of axioms. That being the case, there is every justification for choosing the most complicated form. It is as if Euclid purposely nudged his colleagues: do not cherish any vain hopes, do not seek consolation in the pleasanter equivalents of my postulate, do not attempt to hide the blemish. You will never attain the desired self-evident nature that we require of axioms. This postulate is nothing other than the converse of the parallel theorem! It has to be proved with the aid of the other postulates, or the beauty and harmony of geometry will be ruined. I could not demote this postulate to the rank of a theorem. You try.

To put it briefly, I presume that Euclid had a more profound grasp of the situation than did most of his commentators. Either they were hypnotized by their own analyses and convinced themselves that the postulate was proved, or they attempted to formulate some equivalent and "more natural" postulate. Now Euclid most likely clearly understood that he had not been able to resolve the first problem, and to seek self-evident statements would mean simply to aggravate the illness.

In this rather balanced version of matters there is a weak spot (of course). If there were some kind of investigations, then why didn't Euclid publish them? That is not clear to me. Possibly he felt some inconvenience in putting forth theorems that did not lead to any results. Perhaps, like many great scientists, he did not like to make public uncompleted studies. Take Gauss, who did not publish his investigations into non-Euclidean geometry! But maybe there was a manuscript after all.

That is my strong point: there is very little information to prove or disprove anything in this matter.

Actually the best source of antiquity on the history of the fifth postulate is Proclus' commentary on Euclid. This, as the reader should bear in mind, was in the fifth century AD.

Here we take leave of Euclid. In parting, allow me to say a few warm words.

Euclid was a brilliant mathematician. He was a great teacher. One wants to believe that he was just as good a man and that he lived a long and happy life in his sunny Alexandria, drinking with friends the sweet wine of Chios or the pungent wine of Cyprus—diluted of course because inebriation is a sin of the Scythians but not of the Greeks—joking tolerantly about Ptolemy, instructing his pupils, reading Homer and working to the very end of his days. We hope that he praised the gods of Olympus for making him a geometer.

That is an acceptable way of thinking, and since, for lack of facts, no one can disprove it, we will stay with it.

And with that, farewell to you, Euclid.

The problem has been posed.

Let us see what happened then.

But first...

The Appendix that I promised. A List of the axioms of plane geometry

Six Basic Concepts are introduced. Namely, three Basic Terms: point, line, plane and three Basic Relations: containing (incidence), lying between (for points), motion.

I. Axioms of connection

1. *One and only one straight line can be drawn through two points.*

2. *A straight line contains at least two points.*

3. *There exist at least three points not located on one straight line.*

II. Axioms of order

1. *Among any three points of a straight line, there is always one and only one that lies between the other two.*

2. *If A and B are distinct points of a straight line then there is at least one point C that lies between A and B.*

3. *If a line intersects one side of a triangle (that is, contains a point lying between two vertices), then it either passes through the vertex of the opposite angle, or intersects another side of the triangle.*

By employing the axioms of order it is possible to define very important notions that will be needed later on. Namely: the concepts of *line segment*, *half-line* (or ray), and *angle*.

III. Axioms of motion

For mathematicians, motion is a basic (primary) concept. The properties of mathematical motion are defined by the following axioms.

1. *For a given transformation of motion (call it $\mathcal{M}$) any point A of the plane undergoing transformation passes into a single definite point A'.*

2. *For any transformation of motion $\mathcal{M}$ and any point A', there exists a certain point A that passes into A'.*

3. *For a given transformation of motion $\mathcal{M}$, distinct points A and B are carried into distinct points A' and B'.*

These three axioms demonstrate that motion is a one-to-one transformation of a plane into itself.

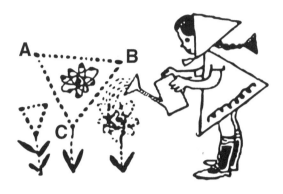

4. *A sequential execution (composition) of any two transformations of motion $\mathcal{M}_1$ and $\mathcal{M}_2$ is also a transformation of motion. We shall denote it $\mathcal{M}_2 \circ \mathcal{M}_1$.*

5. *Every motion $\mathcal{M}$ has an inverse motion $\mathcal{M}^{-1}$, such that the product $\mathcal{M}^{-1} \circ \mathcal{M}$ is a motion that leaves all points of the plane unchanged, that is, it is a so-called identical transformation.*

In view of Axiom 4 it is obvious that an identical transformation should be regarded as a special case of the transformation of motion.

These are followed by axioms which demonstrate that motion does not lead to any "deformation" of the plane.

6. *If a motion transforms the ends of a line segment AB into the ends of the line segment $A'B'$, then any interior point of AB is carried into an interior point of $A'B'$.*

Now comes a most important axiom, without which it would be impossible to establish the concept of the *congruence* of figures.

7. *If A, B and C are three points of some figure that do not lie on a single straight line, then this figure may be moved so that:*

(a) point A coincides with any preassigned point A' of the plane;

(b) the ray AB coincides with any preassigned ray $A'B'$ emanating from point A';

(c) point C coincides with some point C' in any preassigned half-plane resting on the ray $A'B'$ (there are naturally two such half-planes).

Following this, no further movement of the figure is possible.

And, finally, an axiom which shows that mirror reflections are a special case of the transformation of motion.

8. *There are motions that carry segment AB into BA and angle AOB into angle BOA.*

These eight axioms define all the properties of motion, and it is now possible to introduce rigorously the notion of the equality, or, to be scientific, the congruence of figures:

Definition: *Figure S is congruent to figure S′ if it can be made to coincide with figure S′ by means of motion.*

It is now easy to prove the following theorems:

1. Figure S is congruent to itself.

2. If S is congruent to $S′$, then $S′$ is also congruent to S.

3. If S is congruent to $S′$, and $S′$ is congruent to $S″$, then S is congruent to $S″$.

The axioms of plane geometry are nearly exhausted.

What remains are:

IV. The axiom of continuity (Dedekind's axiom).

If all the points of a straight line are partitioned into two sets—I and II—such that any point of set II lies to the right of any point of set I, then either in set I there is a rightmost point and then in set II there is no leftmost point, or conversely, set II has a leftmost point, and then set II does not have a rightmost point.

To put it crudely, this axiom implies that there are no gaps or empty spots in the straight line. It is necessary to introduce this axiom so as to be able to construct a rigorous theory for measuring line segments.

And finally:

V. The parallel axiom

Only one line can be drawn parallel to a given line through a given point not on this line.

We might jump ahead in our story for a moment to say that the axiomatics of Lobachevsky's geometry differs from Euclidean axiomatics solely in this last axiom. All the other axioms of both geometries coincide.

Chapter 4

The Age of Proofs. The Beginning

We begin with a short list of names. The problem of parallel lines was attacked by Aristotle, Poseidonius, Ptolemy, Proclus, Simplicius, and Aganis in the ancient world; by al-Hasan, al-Tusi, al-Shanni, al-Nayrizi, Omar Khayyam, Ibn al-Haytham in the East.

By Clavius, Wallis, Leibniz, Descartes, Playfair, Lagrange, Saccheri, Legendre, Lambert, Bertrand, Fourier, Ampère, d'Alembert, Schweikart, Taurinus, Jacobi in Europe.

And by scores of known and several thousands of nameless mathematicians as well.

The problem of the fifth postulate wrecked so many minds it would be possible to fill a goodsized psychiatric hospital.

That is no exaggeration either. Many spent their whole life in vain attempts at a proof, winding up in mystical terror or a psychiatric ward.

One of the most unexpected indications of the exceeding popularity of the problem lies in a remark made by St. Thomas Aquinas.

St. Thomas Aquinas was a most prominent theologian of the christian world. In one of his researches he found it necessary to solve a problem of exceptional difficulty: "What is beyond the capability of God?"

He pointed out a number of items in this class.

According to St. Thomas Aquinas, God cannot drastically upset the fundamental laws of nature. For instance, he cannot turn a human being into a donkey. (It might be worth adding that most people handle that problem daily without any divine aid.)

To continue, God cannot tire, be angry, sad or take away man's soul, and the like.

The list also contains an item that states that God cannot make the sum of the angles of a triangle less than two right angles.

I am almost convinced that this example is not accidental. St. Thomas Aquinas could have chosen any other more self-evident theorem. It is very likely that he chose this one for the simple reason that he was familiar with vain attempts to prove the fifth postulate and with the fact that the assertion that "the sum of the angles of a triangle is equal to two right angles" is equivalent to the fifth postulate.

It is ordinarily supposed that this theorem became known in Europe in the 18th century. St. Thomas Aquinas lived in the 13th century.

But we must also say that Arabian mathematicians fundamentally investigated the problem of parallel lines and, among other things, obtained that result as well.

Many works might have been known in the early Middle Ages that were subsequently lost.

Today it is hard to realize just how hopelessly confused was the whole theory of parallel lines prior to Lobachevsky.

Today any good mathematics major in college would need no more than two or three weeks of calm work to prove the theorem: if the sum of angles of a triangle is equal to π, then the fifth postulate holds.

And he would prove it even if he were almost totally unfamiliar with non-Euclidean geometry and, consequently, formally in the same position as geometers of the past.

As recently as the 18th century this theorem was considered—and rightly so—to be an important scientific attainment. I do not in the least wish to defend the obviously pleasant thesis that "people are more talented today". That is not the point at all. Simply, in scientific work, confidence in the ultimate result, a clear-cut knowledge that the approach is correct proves to be an almost decisive factor.

An American physicist is reported to have said that as soon as the atomic bomb was exploded the production of it ceased to be a secret. This may be a slight exaggeration, but in principle it is correct.

I am sure the reader will recall how much simpler it is to solve a problem or to prove a theorem if the answer is already known.

Now in the whole problem of parallel lines, only one guiding idea is needed: the fifth postulate of Euclid is independent of all the others. With just that knowledge, any mathematician today would readily repeat most of Lobachevsky's results in a very short time. But he would remain an ordinary mathematician. He would simply know one thing: "you have to dig here." And that would solve almost everything.

I think a case from chess can offer enough supporting evidence. Take any chess puzzle which states that the white has a winning move. The usual requirement in such a position is to find an elegant combination of moves. Any decent chess-player can resolve 90 per cent of such problems in an hour or so. Yet in 90 cases out of a hundred he would never see such a combination in an actual game.

These remarks are to forestall any stupid feelings of superiority over mathematicians of earlier ages. It is true that most of the theorems related to the attempts to prove the fifth postulate are quite elementary in their logic, and quite accessible to grade-school students. What is more, the logical errors of those who thought they had proved it are also very elementary. But their elementary nature is evident only today. In the very same fashion, twenty years hence many of the problems that plague scientists nowadays will appear ridiculously simple and naive. That is what so often happens in physics.

After this heavy dose of general discussion, it is high time to return to the fifth postulate.

I have time and again repeated (the reader will have to excuse me—I admit I'll have to do so again), that all attempts at a proof were motivated actually by a single factor: a certain lack of elegance, a lack of beauty, as the artist would say.

It rankled and it ruffled the aesthetic feelings of scholars by its complexity. The reaction to it was the same in ancient Greece, in Persia and in Europe.

How delightful was the indignation of one of the greatest mathematicians of the Arabic world, Omar Khayyam.

"... Euclid thought that the reason for the intersection of straight lines was that the two angles [interior angles on one side] are less than two right angles.

"In so believing he was right, but it can be proved only with the aid of supplementary arguments. [Khayyam believed that he had proved the fifth postulate]... But Euclid accepted this premise and proceeded from it without proof. I swear upon my life... that here we rightly need the aid of reason..."

"How could Euclid have permitted himself to enter this statement in the introduction [which means choosing it as an axiom] whereas he proved far more simple facts..."

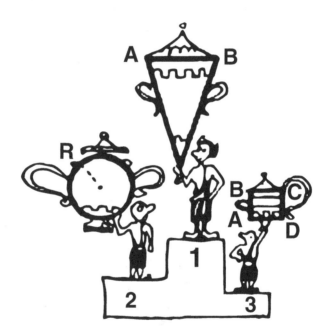

Let us see how the struggle went with the fifth postulate. There were three canonical approaches.

1. A postulate equivalent to the Euclidean one was openly proposed. These authors formed a group called the "modest" or "pessimistic" trend.

2. *Reductio ad absurdum* is one of the most elegant and powerful of logical methods of solving mathematical problems. Here, no new postulates were introduced.

A theorem contrary in meaning to the fifth postulate or to one of its equivalents was formulated; this was followed by the elaboration of diversified corollaries in the hope that sooner or later all this would lead to a contradiction, which would *ipso facto* prove that the fifth postulate followed from the other axioms, and the problem would be solved.

This is the optimistic, presumptuous trend.

3. And, finally, we have the group of "eclectics".

They proved some theorem equivalent to the fifth postulate. And they proved it with the unwitting employment of some other equivalent of Euclid's postulate.

Trend No. 2, the optimists, had the hardest time. They kept stringing out the chain of their theorems, floundering more and more in the corollaries, and still finding no contradictions.

From the vantage point of today we realize that this group of mathematicians actually were proving the initial theorems of non-Euclidean geometry, and that they were on the most promising pathway, for only in this way one could come to realize that the Euclidean postulate was independent of all the others. But that did not make things easier for them. As a rule, they either lost heart or went over to the camp of "eclectics".

One must note that many of the proofs of the eclectic group are magnificently witty.

To somewhat simplify the actual history, one might say that, in the main, attempts were made to prove two basic varieties of the fifth postulate:

1. A perpendicular line and an inclined line meet.

2. The sum of the angles of a triangle is equal to π.

In this way, several very pictorial equivalents of the fifth postulate were found. At times the authors realized that they had found an equivalent; at other times they were deluded into thinking that they had proved the fifth.

Here are a few ersatz postulates:[1]

1. The locus of points equidistant from a given straight line is a straight line.

2. The distance between two nonintersecting straight lines remains bounded.[2]

3. There exist similar figures.

4. If the distance between two straight lines first diminishes upon motion in some direction along these straight lines, it cannot begin to increase until the straight lines meet.

---

[1] In formulating the equivalents of the fifth postulate, I will always presume that everything occurs in one plane.

[2] This is a less strict demand than No. 1.

And so on.

In all there are about thirty such statements.

For the amusement of the reader, I give several "proofs" of the fifth postulate without any critical commentary. If he so desires, he will be able to figure out what postulate was used each time in place of the fifth.

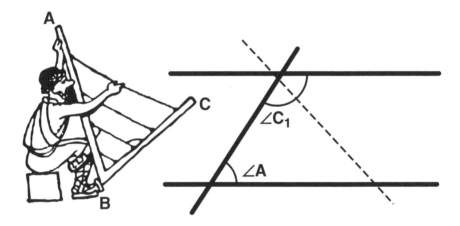

1. The proof of Proclus. One of the very first, one of the simplest and one of the cleverest.

Proclus starts out with Aristotle's assertion: When we go along two straight lines from a point of intersection, the distance between them increases without bound.

He takes this to be an axiom.

Actually, it is a theorem. What is more, it is a theorem that is quite independent of the fifth postulate. So we can rely fully on this theorem. It belongs to "absolute geometry" and, hence, as we understand matters today, it holds true both in Euclidean geometry and in the geometry of Lobachevsky. But the postulate—its equivalent used implicitly by Proclus—is different.

Here is the proof, actually an outline of the proof. (I will not hold rigorously to the formal schemes of this and the following proof. That would be a little too much.)

Draw two definitely parallel straight lines; that is, such that $\angle A + \angle C_1 = \pi$.

Draw a third straight line. How? It is shown in the figure as a dashed line.

The distance between the dashed line and the upper line (when moving rightwards) increases without bound. Consequently, there will come a time when it will exceed the distance between the parallel lines.

And it will be clear now that the dashed line will cut the lower line.

We suggest the reader to formulate matters in rigorous form and also state which postulate Proclus employed implicitly.

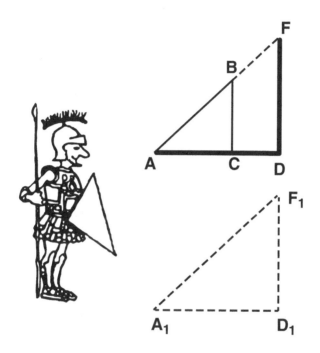

2. The proof of Wallis

We will prove that a perpendicular and an inclined line to a common transversal intersect.

From point B drop a perpendicular to the transversal, producing a triangle ABC. Take a similar triangle such that its side corresponding to AC is equal to AD.

In view of its importance, draw this triangle ($\triangle A_1 D_1 F_1$) separately.

Superimpose the dashed triangle on our $\triangle ABC$ so that $A_1 D_1$ lies on AC. Then the side $A_1 F_1$ will lie on our inclined line and the side $D_1 F_1$ will lie on our perpendicular.

Actually, the proof is complete: there are only a few formalities left. I leave them to the reader.

Let us not get too involved in examples. A more interesting point is the following. Dozens of mathematicians, people of diverse cultures, separated by centuries of time and frequently not even knowing the existence of one another reasoned in almost identical fashion, repeating one another in almost the same words.

Prior to the 18th century, proofs of the fifth postulate via the method of *reductio ad absurdum* did not string out the chain of corollaries too far and did not delve deeply in analysis. At some moment people would say: "Behold! There is the contradiction!" But actually the contradiction was not with the axioms of the "absolute geometry", but with an implicitly introduced equivalent of the fifth postulate.

But since matters did not go very far, there were more hunters than rabbits. There were more mathematicians working on the fifth postulate than there were distinct modes of alleged proof. Almost all the greatest mathematicians of the world engaged the fifth postulate.

There is one about whom I want to say a bit more. Not because his investigations into the theory of parallel lines were something exceptional. No, not at all. His most interesting results were obtained in the field of algebra. He did not advance much beyond any of the others in the theory of parallels. In this sense we will be giving him more than his due of attention. What is more, we will not even speak much about his proof of the fifth postulate. True, the proof he offers is extremely clever. True, again, his influence on subsequent studies of Oriental mathematicians is definitely felt. Finally, the technique which he employed was extremely suitable and was in advance of Western mathematicians by six hundred years. (We shall touch on this matter somewhat later.) But, in fact, it is not the fifth postulate that interests us so much in this book.

The exciting thing about this man is that he is a case which illustrates beautifully how little are the differences between people of all nations and all ages.

I now speak of the mathematician that is known as the poet Omar Khayyam.

Chapter 5

Omar Khayyam

The full name is Ghiyath al-Din Abu'l Fath Omar ibn Ibrahim al-Khayyami an-Naisaburi. To Europeans he is simply Omar Khayyam.

The East, as we all know, is the East, in contradistinction to the West, which is the West.

The East, the Orient—to most people it signifies the usual collection of harems, sultans, the Islam, califs, hookahs, emirs, minarets, shahs, muezzins, burning sun, fountains, Genghis Khan, houris, and the shade of sycamores. The stifling heat of the high-noon sun, and lazing in the shade.

That is the Orient of the past, at least, the way some people picture it.

All these things—sultans, califs, emirs and the rest—existed in the East. Moreover, many of them are still found there in some parts.

Notwithstanding, there never was any East.

There were and still are dozens of countries and over a thousand million human beings. These millions upon millions of people are quite diversified.

One might presume that their inner world is the same as that of dwellers in the West.

Incidentally, Kipling, who coined the famous phrase about the East and the West, thought so. Such is the idea that is advanced in his celebrated ballad, of which people usually remember only the first line (such, alas, is the fate of many a brilliant poet).

Since this chapter will be permeated with the "atmosphere of poetry", let us take a few lines of Kipling's poem, all the more so that they are indeed beautiful lines.

> Oh, East is East, and West is West, and never the twain shall meet,
> Till Earth and Sky stand presently at God's great Judgment Seat;
> But there is neither East nor West, Border nor Breed, nor Birth,
> When two strong men stand face to face, tho' they come from the ends
> of the earth!

There is no use quoting any further because what follows is pitifully bad. The poetry is still excellent, but the topic and its resolution is a terrible let-down, hardly better than a routine Hollywood film about the Wild West.

Kipling confined himself to a hymn in honour of the spiritual unity of warriors, heroes strong in body and spirit. Taken at face-value, these warriors are something in the nature of a preimage of the noble bandits of Hollywood. But if one ignores his choice of heroes, he can fully agree with Kipling. Gangsters throughout the world find a common language with just as much ease as humanists in the world of science.

Unfortunately, Kipling sang the praises of the former and gave to them his amazing poetical talent.

This whole discussion is very much to the point if one recalls that we are speaking of Ghiyath al-Din Abu'l Fath Omar ibn Ibrahim al-Khayyami an-Naisaburi.

Ghiyath al-Din means "the help of faith" and is a traditional title for all scholars, since in those days the hierarchical ladder of scientific knowledge was apparently not so involved. Abu'l Fath means the father of Fath.

An-Naisaburi indicates that Khayyam was born in Nishapur, which was one of the chief cities of glorious Khorassan.

Khayyam—what we have taken as the last name—means tent-maker. Most likely his father or grandfather was so engaged.

Ibn Ibrahim is the son of Ibrahim.

Finally, Omar, is the given name.

In short, Omar Khayyam, who conquered the West in the 19th century and conquered it as a poet.

He was first translated into English and came out in 25 editions last century. In England and America admiration for Khayyam developed into an epidemic. He was quoted and praised, and clubs named after him sprang up everywhere. Willy-nilly we shall have to delve into the literary studies, and I want to say right away that though his poetry is indeed beautiful, but his exceptional popularity is due possibly to a certain "marvellous revelation". It turned out that a thousand years ago, somewhere in Turkey, or India, there lived people whose thoughts and emotions did not differ much from those that excited people living in the civilized age of the 19th century. More, Khayyam casted these thoughts and emotions in magnificent poetical form, which was double amazing.

On the other hand, in his home land he was hardly at all known as a poet.

Thus arose two Khayyams.

In the West was the poet.

In the East, the mathematician, astronomer and philosopher. Oh, East is East and West is West.

Then who is this Omar Khayyam?

Since I lean more to the oriental version, let us begin our story of the honourable wise man and imam Omar al-Khayyami of Nishapur, may Allah sanctify his dear soul.

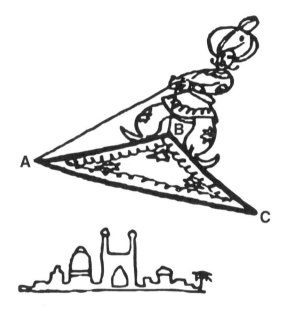

"In the name of the gracious and merciful Allah, praise Allah, the lord of the worlds, and blessing unto all his prophets."

Thus did Khayyam, bound by a rigid traditional form, begin his marvellous *Treatise on the Proofs of Problems of Algebra and Almuqabala*, a mathematical work that was roughly five hundred years in advance of the mathematics of the Occident.

This work of the "greatest geometer of the East", as a remarkable encyclopaedist of the Orient Ibn Khaldun wrote of him later, contains the first systematic theory of third-degree algebraic equations. It was well known to Arabian mathematicians and undoubtedly exerted a tremendous effect on the development of mathematics in the East. In Europe, the first and rather nebulous reference to it occurs only in the year 1742.

Ibn Haldun actually only says that it would seem, by the title of the manuscript, which is in the Leyden Museum, that one may suspect that it contains something about equations of the third degree, but "It is such a pity that none of those who know Arabic has any taste for mathematics and none of those who have mastered mathematics has any taste for Arabian literature."

When the treatise of Khayyam was finally read, it was found that his results were repeated (and in many respects surpassed) by no other than Descartes. Incidentally, it is possible that in yet another treatise that has been lost irretrievably Omar Khayyam went much farther. Who knows?

But we are more interested here in another treatise of Omar Khayyam, to wit: *Commentaries on the Difficulties in the Introductions to the Books of Euclid.* This composition of the most glorious sheikh, imam, of the Proof of Truth, of Abu'l Fath Omar ibn Ibrahim al-Khayyami is in three books.

Again, this treatise, in the beginning, lacks originality: "In the name of Allah, so gracious and merciful, Praise Allah, the lord of grace and mercy, and peace be unto his slaves and in particular unto Muhammad, the lord of the prophets, and unto all his pure clan."

All this ritual breaks off suddenly just a line down: "The study of the sciences and the comprehension of them by means of true proofs is necessary for him who seeks salvation and happiness."

That's enough. He who was eager to understand, did. Already too much was said. On went the soul-salvaging ritual.

"And especially [of course, most naturally] this refers to the general notions and laws to which one resorts in studies of the hereafter, proof of the existence of the soul and its eternalness, comprehension of the qualities that are necessary for the existence of the Almighty and his magnificence [Khayyam is worried beyond reason about the magnificence of Allah], the angels, the order of creation and proof of the prophecies of the lord, the prophet [Muhammad, that is], to the orders and prohibitions of which bow in obedience all creatures [incidentally, there was a time—in Medina—when Muhammad introduced a very rigorous order and the best of the creatures of Allah were on their toes] in accord with the pleasure of the Almighty Allah and the power of man."

What a flawless piece of writing, it would seem.

Yes, it would seem, for the entire paragraph is one solid heresy, extremely dangerous to any orthodox preacher of Islam.

And let the worshiper of Aristotle smooth over his writing with hypocritically pious phrases. He will be understood by those of the same views, but by those of the opposite views as well.

Omar's luck that, in general, Islam was a more tolerant religion than Christianity. On the average, that is. There was no burning at the stake. But one could expect, when needed, a swift plunge of the dagger. Very much so, in fact. Even for just a tiny bit of heresy. On the other hand, one could get around that too.

Then follows the treatise proper. (We shall have more to say about it later on.) All the way along, however, Omar put in the proper proportions of glory to the Almighty Allah, and to his greatest creation, Muhammad, and to the whole lineage of Muhammad, to the great help of Allah and more and more.

Praise Allah!

How merry and nice it was for his creations. His creatures I mean. Noting that the merciful servants of the merciful Christ made the life of Lord's creatures under their controle even merrier, we begin again "in the name of the gracious and merciful Allah".

We know hardly anything at all about Omar Khayyam, only a few fragmentary bits here and there. By way of complicated "astronomical" computations on the basis of indirect findings, the dates of his life are, approximately, fixed from 1048 to 1131. Or from 1040 to 1122. Or from 1048 to 1122.

He was born in Nishapur. At that time, the city was located in the emirate of Khorassan. Today, Nishapur is in the north east of Iran. Omar wrote his verses in the literary Persian language, and his learned studies in Arabic. Since, as linguists explain, both modern Persian and Tajik developed out of medieval Persian, we may justifiably say, today, that Khayyam is a Persian poet and a Tajik poet.

A few years prior to the birth of Omar Khayyam, that region of the "calm and lazing" Orient was the scene of bitter battles, and the leaders of the nomad Seljuks (Turkmens) first routed the earlier sultans and then set up a collosal empire and a nice fresh dynasty of Seljukian sultans.

What followed was rather standard. Fighting for the throne among the aspirants. The sultans fighting feudal lords and frenzied attempts of the feudals to rule by themselves, independently. In about one hundred and twenty years the empire fell to pieces completely. But that period of time, which to history is minuscule, to a human being is quite enough.

Khayyam lived in the empire of the Seljuks and lived quietly for a long time, for he had a patron. A strong protector.

The great vizier Nizam al-Mulk.

Nizam al-Mulk was possessed with the idea of a strong state. And he furthered it in many ways. He apparently believed that culture and learning would strengthen his empire and so, like those dear Ptolemys of antiquity, he patronized his scholars in many a way.

He himself was not above literary forms and wrote a rather serious, fundamental and very interesting—to historians—work entitled the *Book of Government*—a sort of handbook for sultans who needed training (they certainly did). In this work of popularization he engaged the services of his scholars and in particular those of Omar Khayyam.

But before Omar entered into the service of Nizam al-Mulk he had endured much indeed.

When a sultan is setting up an empire, the inhabitants do not have it easy at all.

There is practically no information about the youth of Omar, other than that he may have studied in Nishapur.

The story goes that "at the age of seventeen years he attained profound knowledge in all fields of philosophy".

It is said that he was "a profoundly knowledgeable man in linguistics, Muslim law and history" and was a follower of Avicenna (Abu Ali ibn Sina).

It is also related that he had a marvellous memory and that on one occasion he learned a whole book by heart after reading it seven times.

Some said that he was a "sage with extensive knowledge in all fields of philosophy, especially mathematics".

In a word, then, all sources (and also the writings of Khayyam himself) describe a man with encyclopaedic knowledge and a mind of exceptional gifts and perspicacity.

At the beginning, however, all these good points worked more against him than for him. He was compelled to leave Khorassan, and we find Omar Khayyam in Samarkand.

Quite naturally, a patron was needed. And Omar found him. We do not know how, but he did. This "marvellous and incomparable judge of judges the imam Abu Tahir, may Allah continue his rise and may Allah cast aside those who are envious and wish him evil".

To put it simply, this was the chief judge of Samarkand, a high-placed official. But only Allah really knows whether he possessed even a minute portion of the merits that Omar so painstakingly and sweet-singingly described in his algebraic treatise. A bit earlier, in the introduction to the same treatise, Omar wrote sullenly and bitterly:

"... I was deprived of the opportunity of engaging regularly in my studies [of algebra] and I could not even concentrate on meditation about it because of the reverses of destiny that plagued me.

"We were witness to the death of learned men, of whom there remains a small and suffering group. The harshness of fate in these times prevents them from giving themselves wholly to refining and deepening their learning.

"Most of those, who today have the aspect of a scholar, dress truth in falsehood, without going beyond imitation in science and only pretending to have knowledge.

"The knowledge which they have amassed is used for base purposes of the flesh. If they encounter a man that seeks the truth and loves the truth, if he attempts to reject falsehood and hypocrisy and give up boasting and deceit, they make him the object of their contempt and mockery."

When reading an excerpt like this, one no longer wishes to relate the story of Omar Khayyam in the cool and slightly ironical tone of the objective observer. There is no place for words about the great and merciful Allah. Here life is harsh and cold. Note that these bitter lines were written by a very young man. At that time he was hardly more than twenty-five years of age. Such a desire vanishes completely when we recall that four centuries later almost the very same thing would be written by Galilei, and within another five centuries by Einstein.

I am not sure what Omar wanted to say, but the next sentence ("Allah helps us in all cases, he is our refuge") followed by an extremely long paragraph praising the honourable judge of Samarkand reads like a savage, vicious, razor-sharp taunt.

Let us not stray. Omar was lucky. He found a patron. What is more, one "whose... presence opened up my chest and whose society elevated my glory, my work expanded due to his light and my back was strengthened because of his good deeds".

So you see how wonderful everything was. Yet that was only the beginning. Allah is never grudging in his generosity.

Omar Khayyam is honoured (glory be unto Allah!) by the friendship of the khakan of Bukhara himself. What this title signifies, I do not know, nor have I tried to find out. At any rate, he was some kind of minor king of sorts. And an historian (a contemporary of Khayyam) reports, with an understandable tinge of envy, that "... khakan Shams al-Mulk elevated him greatly and seated the imam Omar on his throne".

But the good deeds of Allah are indeed inexhaustible. And in the year

1074 Malik Shah himself (the khakan is only a vassal of the shah) summoned Omar to his court in Isfahan and—rejoice oh ye faithful—makes him his nadim.

You would probably like to know what a nadim is.

A rather strange post.

A sultan is always in need of interlocutors, confidants, body-guards. Those are the duties of the nadim. He has his meals with the ruler, converses with him and engages him, thinking up all kinds of things in order to kill time. And of course he shows his admiration for the wisdom of the ruler, the courage, the beauty, the poetical gifts of the sultan, for his steeds, his eagles and his concubines. True, I do not know whether nadims were demonstrated the most beautiful flowers of his harem or not, but....

No need for this amateurish talk, we give the floor to the radiant patron of Omar Khayyam. Nizam al-Mulk himself.

We quote from the *Book of Government* (*Siasset-Nameh*).

"The benefits of the nadim are several: one is that he is a close friend of the sovereign, another is that since he is with the sovereign day and night, he acts as a body-guard, and in case of necessity—do not allow it, oh Allah—if there is some kind of danger, he sacrifices his body by using it as a shield against that danger; and fourthly, a thousand kinds of words can be said to the nadim rather than to those who perform the duties of the ministers and the officials of the sovereign; the fifth benefit is that they report, like spies, on the affairs of the kings; the sixth, that they converse in all manners without compulsion about good and evil, whether inebriate or sober, thus bringing about much that is useful and purposeful."

So you see, six distinct benefits. Few indeed can occupy such an honourable post. Very few.

"It is necessary that the nadim be gifted by nature, virtuous, good-looking, of pure faith, a guardian of secrets, well-mannered; he must be a narrator of stories, a reader of what is merry and what is serious, he must remember many legends, he must always be ready with a good word, a reporter of pleasant news, a player of backgammon and chess, and if he can play some musical instrument and handle arms, all the better. The nadim must be in accord with the sovereign. To everything that takes place or that the sovereign utters, he must answer: 'Excellent, marvellous'; he may not instruct the sovereign with words 'do this, do not do this, why did you do this?' He must not so speak because the sovereign will then be depressed and will reject him. It is proper for the nadim to arrange all matters pertaining to wine, recreation spectacles, friendly congregations, hunting, the playing of chovgan and the like, for that is what they are needed for."

That is all.

Thus preached Nizam al-Mulk, who presented Khayyam to Malik Shah as nadim.

Without doubt, an amazingly pleasant post.

Historians console us somewhat. One group thinks it highly improbable that Khayyam was honoured so greatly, and they believe that the biographer exaggerated. Perhaps he wished to elevate a scholarly colleague in the eyes of the reader and allowed for some exaggeration, a bit of boasting. Others feel that Khayyam was indeed a nadim but, they say, of a somewhat different kind.

For Nizam al-Mulk continues: "Many sovereigns have made physicians and astrologers their nadims so as to know the opinion of each of them as to what should be done by them, what by the sovereign, what needs to be done to preserve nature and the health of the sovereign. Astrologers observe the time and the hour; for every matter that is pleasant they give notice and select a favourable hour."

In general, then, there is a faint hope that Khayyam did not have to arrange the drinking sprees of Malik Shah and locate concubines for him. But who knows? One thing we can be sure of is that he had to do everything that came into the head of the ruler.

At any rate, he definitely delved into astrology, though just as definitely he believed it to be nonsense.

As astrologer, Omar Khayyam was an indisputable authority, but it is his secret how it came about.

And with what professional skill one had to cringe in the courts of the East! Times without number!

On the whole, then, this life which to many was so pleasant, was thoroughly disgusting to Omar Khayyam.

There were a few things in exchange, though.

Firstly, the court sage of Malik Shah, his confidential agent, almost pal, was inaccessible to all servants of the Koran, who were oh so eager to make Omar toe the line.

Secondly, Omar was well provided for. True, he did not have a family, but the position of a scholar in those days was precarious, so much so in fact that it was impossible to exist without a patron. So better a shah than some kind of small fry.

Thirdly, perhaps most important of all was the possibility to work. Omar had at his disposal what at that time was a first-class scientific institution, the Isfahan Observatory. And probably the shah reasonably assumed that his wiseman should have some spare time for meditation. At any rate, Khayyam did a great deal during his years at court. Three years after his arrival here he had already completed his *Commentaries to the Difficulties in the Introductions to the Book of Euclid*, where among certain other corrections he proved—so he thought—the fifth postulate.

He was busy in the observatory and obtained excellent results. Actually, he was the first director of the observatory, for the building of which he constantly requested money of Malik Shah.

Again a routine situation.

Nobody was interested in his astronomical work. He compiled a calendar that was marvellously accurate, but the calendar was never accepted. On the other hand, his astrological studies were viewed as undoubtedly valuable.

A number of centuries later, Kepler, who valued astrology like Khayyam did, tread the same pathway. It was solely through astrology that he achieved a position in society, the means for daily life and the opportunity to engage in scientific studies.

Omar did not believe in astrology. Historians have not yet decided what his belief was. There seems to be one and perhaps the most important symbol of his faith: a person should study science and learn about how the world is made. But here too the situation is complicated, so it is time to return to his verses. Speaking generally, if we knew exactly which verses were indeed written by Omar, they would be an extremely valuable document.

He did not consider himself a poet. Most likely he wrote for himself and was naturally less secretive than in his philosophical treatises, in which he always had to be extremely careful, cautiously interpolating minuscule deviations from orthodoxy. Meanwhile specialists in literature are still fighting over which verses are genuinely his.

The canonical text is claimed to contain 252 rubais (quatrains). But here too the debate continues. A total of about 1,000 quatrains are ascribed to Omar.

We shall take it that the verses are genuine. Nevertheless it is rather difficult to determine exactly the philosophical world view of Omar Khayyam. Even the specialists are unable to reach a single opinion, which, incidentally, is how things usually stand.

Some of the verses are magnificent even in translation, and are better in the original, so they say. True, Omar's topics are rather restricted; frankly speaking, twenty to thirty verses fully exhaust everything that Khayyam wanted to say.

Now, so that the reader can rest with some good prose and nice poetry, I will quote a somewhat unorthodox analysis of Khayyam's works and then a few of his quatrains.

O. Henry, probably irritated by the Omar Khayyam craze, put the matter in story form as follows.

The hero of his story *The Handbook of Hymen*, Sanderson Pratt, cowboy, was caught in a snow storm in the mountains and had to sit it out with another cowboy, Idaho. It was most likely a case of psychological incompatibility: tragedy was averted when they found two books.

One was a "Handbook of Indispensable Information" and the other was Omar Khayyam. In a card game, Idaho won and chose Khayyam, and Sanderson got the handbook. Over the monotonous weeks each studied his book.

At last released from the snow, the two cowboys returned to normal life and began paying court to a charming wealthy widow, each displaying his newly acquired culture and utilizing to the fullest what he had read. Idaho's poetic guide—Omar Khayyam—was defeated roundly by the handbook, and the happy marriage of Sanderson Pratt was the worthy reward of the bearer of common sense. As Sanderson Pratt put it:

"I sat and read that book for four hours. All the wonders of education was compressed in it. I forgot the snow, and I forgot that me and old Idaho was on the outs. He was sitting still on a stool reading away with a kind of partly soft and partly mysterious look shining through his tanbark whiskers.

" 'Idaho,' says I, 'what kind of a book is yours?'

"Idaho must have forgot our quarrel, too, for he answered moderate, without any slander or malignity.

" 'Why,' says he, 'this here seems to be a volume of Homer K. M.'

" 'Homer K. M. what?' I asks.

" 'Why, just Homer K. M.,' says he.

" 'You're a liar,' says I, a little riled that Idaho should try to put me up a tree. 'No man is going round signing books with his initials. If it's Homer K. M. Spopendyke, or Homer K. M. McSweeney, or Homer K. M. Jones, why don't you say so like a man instead of biting off the end of it like a calf chewing off the tail of a shirt on a clothes-line?'

" 'I put it to you straight, Sandy,' says Idaho, quiet. 'It's a poem book,' says he, 'by Homer K. M. I couldn't get colour out of it at first, but there's a view if you follow it up. I wouldn't have missed this book for a pair of red blankets.' "

The new Omar Khayyam convert, Idaho, then gives an analysis of the poet.

" '... He seems to be a kind of a wine agent. His regular toast is 'nothing doing', and he seems to have a grouch, but he keeps it so well lubricated with booze that his worst kicks sound like an invitation to split a quart. But it's poetry,' says Idaho, 'and I have sensations of scorn for that truck of yours that tries to convey sense in feet and inches. When it comes to explaining the instinct of philosophy through the art of nature, old K. M. has got your man beat by drills, rows, paragraphs, chest measurement, and average annual rainfall.' "

But Sanderson Pratt wasn't one to give in easily.

"This Homer K. M., from what leaked out of his libretto through Idaho, seemed to me to be a kind of a dog who looked at life like it was a tin can tied to his tail. After running himself half to death, he sits down, hangs his tongue out, and looks at the can and says:

'Oh, well, since we can't shake the growler, let's get it filled at the corner, and all have a drink on me.'

"Besides that, it seems he was a Persian; and I never hear of Persia producing anything worth mentioning unless it was Turkish rugs and Maltese cats."

Though lovers of Omar Khayyam will be indignant, we must admit that the basic topic was grasped rather neatly by the two cowboys.

True, one never knows what exactly O. Henry is driving at.

It might well be that as a true admirer of Omar Khayyam, he simply wished to illustrate the ancient but sad theme: forget poetry if you want to achieve success with a charming lady, forget or give up all hope. Especially if the woman is the owner of a two-storey house in a neat little provincial town.

Now for Omar Khayyam's verses. Crudely, we might divide them into three groups: (1) the love and wine cycle; (2) the philosophical cycle; and (3) civic lyrics, the quatrains in which Omar describes more or less straightforwardly his attitude towards his surroundings.

Since I have been constantly balancing on the brink of solving psychological enigmas, let us this time try to figure out to what extent Omar's verses convey the true image of the writer himself.

Perhaps in this sense the most meaningful are the verses of the third cycle: irritated, full of gall, definitely vicious.

Of all 252 verses, there is not a single one that says something decent about the thinking creations of Allah. Everyone gets it in the neck, but Omar is particularly bitter towards the clergy.[1]

---

[1] When comparing the Russian original with the translation of George Yankovsky, I was surprised to find out that the English versions of the ruba'is that Yankovsky gave did

Why, all the Saints and Sages who discuss'd
Of the Two Worlds so wisely—they are thrust
Like foolish Prophets forth; their Words to Scorn
Are scatter'd, and their Mouths are stopt with Dust.
(Fitzerald)

And do you think that unto such as you;
A maggot-minded, starved, fanatic crew:
God gave the secret, and denied it me?
Well, well, what matters it! Believe that, too.
(Le Gallienne)

It is quite natural now to go on to the merciful Allah himself. Omar does not get along so well with the Lord in verses as he does in treatises.

---

not coincide with the Russian versions discussed by Voldemar Smilga. Then I looked at the whole corpus of the best poetic translations of Khayyam due to Edward Fitzgerald, consulted Wikipedia, and understood the reason—these translations are rather free and do not strictly follow the Persian original. As a result, it is in most cases difficult to establish an exact correspondence between the two versions. However, with the help of my son Boris Smilga we found on the internet the translations of Edward Whitfield, such that their correspondence with the Russian translations can in many cases be well established. Thus, in order to better illustrate the author's point, many quatrains below will be given in the Whitfield's translations, even though they may not be so good poetically as the translations of Fitzgerald. The third translator whose work we used is Richard Le Gallienne.—A.S.

Who framed the lots of quick and dead but Thou?
Who turns the troublous wheel of heaven but Thou?
Though we are sinful slaves, is it for Thee
To blame us? Who created us but Thou?
(Whitfield)

The Master did himself these vessels frame,
Why should he cast them out to scorn and shame?
If he has made them well, why should he break them?
Yea, though he marred them, *they* are not to blame.
(Whitfield)

And then some general statements concerning men's stupidity. He writes it with taste. One may even say, with appetite.

To wise and worthy men your life devote,
But from the worthless keep your walk remote;
Dare to take poison from a sage's hand,
But from a fool refuse an antidote.
(Whitfield)

In heaven is seen the bull we name Parwin,
Beneath the earth another lurks unseen;
And thus to wisdom's eyes mankind appear
A drove of asses, two great bulls between!
(Whitfield)

This entire cycle may very logically be concluded with quatrains in which Omar explains the situation in which he is compelled to live and work.

When to this loot of life I come anear,
Hoping to snatch some little worldly gear,
I find the fools have carted off the best,
And nought is left for me but hope and fear.
(Le Gallienne)

'Tis written clear within the Book of Fate,
The little always shall oppress the great,
Who most deserves be slave to those who least,
And only fools and rascals go in state.
(Le Gallienne)

How long, how long, in infinite Pursuit
Of This and That endeavour and dispute?

Better be merry with the fruitful Grape
Than sadden after none, or bitter, Fruit.
(Fitzerald)

The writer of such "radiant" verses is definitely not a man with an optimistic turn of mind. Complete spiritual isolation and no breaks in the gloom. In his philosophical ruba'is Khayyam kind of generalises his life experience.

My critics call me a philosopher,
But Allah knows full well they greatly err;
I know not even what I am, much less
Why on this earth I am a sojourner!
(Whitfield)

'Tis all a Chequer-board of Nights and Days
Where Destiny with Men for Pieces plays:
Hither and thither moves, and mates, and slays,
And one by one back in the Closet lays.
(Fitzerald)

Oh, come with old Khayyam, and leave the Wise
To talk; one thing is certain, that Life flies;
One thing is certain, and the Rest is Lies;
The Flower that once has blown for ever dies.
(Le Gallienne)

Again, not a single bright spot, not even a hopeful hint. Still in the first cycle there appear to be certain prescriptions for arranging life. O. Henry's heroes (I can repeat) grasped the gist of the matter quite precisely. Incidentally, the first English translation by Fitzgerald paid special and exceptional attention to this particular trend.

You know, my Friends, how long since in my House
For a new Marriage I did make Carouse:
Divorced old barren Reason from my Bed,
And took the Daughter of the Vine to Spouse.
(Fitzerald)

Did God set grapes a-growing, do you think,
And at the same time make it sin to drink?
Give thanks to Him who foreordained it thus
Surely He loves to hear the glasses clink!
(Le Gallienne)

Ah, fill the Cup:—what boots it to repeat
How Time is slipping underneath our Feet:
Unborn TO-MORROW, and dead YESTERDAY,
Why fret about them if TO-DAY be sweet!
(Fitzerald)

And finally

Ah, with the Grape my fading Life provide,
And wash my Body whence the Life has died,
And lay me, shrouded in the living Leaf,
By some not unfrequented Garden-side.
(Fitzerald)

So many cups of wine will I consume,
Its bouquet shall exhale from out my tomb,
And every one that passes by shall halt,
And reel and stagger with that mighty fume.
(Whitfield)

Not so cheerful, I would say...

For lack of any other hypotheses, let us take it that the writer of these quatrains was indeed Omar Khayyam. At least half of them, that will be enough.

The portrait of the man who wrote these verses would seem to be clear. A clever, gifted skeptic and misanthrope. Definitely cultured, but totally lacking in any kind of intellectual interests, all his days and nights spent with concubines and wine, in the company of drinking revellers; and in a rare sober moment he writes marvellous, but deeply pessimistic verses. He values nothing more in this world than the opportunity to carouse, and does so to the limit of his strength and money.

An indigestible blend of a Byron hero, a low-class patrician of Rome, Goethe's Mephistopheles, the debauchery of a Russian merchant or a French aristocrat.

Omar's ideas are by no means new.

There have been skeptics and pessimists throughout the ages, and their Weltanschauung does not call for admiration.

Omar, at times, appears close to spontaneous materialism. At any rate, he abuses Allah often enough. But, firstly, this point is not so clear—there also are a goodly number of partly mystical quatrains and, secondly, are such abuses so exceptionally interesting? In every age and period of human history, materialistic ideas have inspired many thinkers.

In the case of Khayyam there is no need, however, to make allowance for the intellectual naiveté of that age compared with our own. No need at all to pat past centuries on the back goodnaturedly.

If, however, we speak as equals and judge by verses alone, the image of Omar Khayyam the thinker loses much of its lustre. There remains a magnificent poet, but not a very likeable or profound person. We can understand and justify but we cannot agree.

Literary critics do not speak so frankly, perhaps, because the poetry of Omar Khayyam is firmly placed along with the greats of world culture and so also is Khayyam the man—canonized.

But if I only knew Khayyam the poet, I would, after a period of enthusiasm for his pessimism, between the ages of 15 and 25, agree with O. Henry, though paying full due to his supreme poetical skill.

However, the charm lies in the fact that our hypothetical image is but a caricature, and lopsided at that. Because Khayyam was not a poet by profession. He was a scholar. His business was learning. Verses? Only for recreation.

Houris and wine? If Omar had but imbibed a hundredth part of the wine that flows through his verses. If his harem had contained a tenth of the beauties whose praises he sang—he would not have strength left for anything else.

Yet all his contemporaries—well-wishers and ill-wishers alike—are of one opinion: the hajji imam Omar was one of the greatest of learned men of the East.

He was a—Mathematician. Probably the greatest in oriental history. That, at any rate, is the opinion of many mathematical historians. The algebraic works of Khayyam are—no harm in repeating it—brilliant. He made a thorough study of the mathematical legacy of the Greeks. That in itself is quite some undertaking requiring years of work.

Astronomer. Recall the years he spent setting up the Isfahan Observatory. You remember the constant prolonged astronomical observations he carried out, the reform of the calendar and the newly devised system of chronology.

Part physicist. He produced a highly curious treatise on Archimedes' celebrated problem of Hiero's golden crown, the problem that gave rise to Archimedes' law and the trademark of the *Molodaya Gvardia* (*Young Guard*) Publishing House.

Yet that is not all. From Omar's works it is evident that he had a fundamental knowledge not only of Arabian philosophy but Greek as well, particularly the philosophy of Aristotle. Khayyam was even too openly enravished with Aristotle. This is most evident in the way he refers to Aristotle—briefly and lacking in any emotion. In place of the name, he writes "philosopher".

Philosopher and no oriental compliments. Omar could use epithets when he wanted to. But he didn't here. He did not want embellishments, the inflation of which he felt so keenly; he did not want falsely honeyed phrases to stick to names that were really dear to him.

The *philosopher* was enough.

Generally, when Omar gets down to business, the poetical, courtier, oriental style vanishes without a trace. Between the traditional bows to Allah, Muhammad and the current patron at the beginning and end of each piece of writing, we find a restrained and reserved text.

References, arguments, drawings, formulas. Euclid is simply Euclid and not the prince of mathematicians or the beacon of knowledge. Apollonius is simply Apollonius. Ptolemy just Ptolemy. A touch of editing here and there and the style is that of the twentieth century. And Aristotle is the philosopher.

We have strayed a bit. What is interesting here is something else. Recall that "the philosopher" wrote in a very turgid confused style. Any detailed study of his writings is an exceptionally difficult job. I'm sure that today there are not many specialists in the history of philosophy that have worked through all of Aristotle's legacy in the original Greek. Perhaps only a few scholars specializing in the life and work of Aristotle. Now there is no doubt that Omar Khayyam studied all his works. Yet Aristotle is only a small part of the philosophical legacy of the Occident and the Orient that Omar studied, as is so eloquently witnessed to by references to dozens of diversified fundamental writings.

Speaking of the volume of digested literature, Khayyam is the envy of any academician in philosophical science.

Philosophy does not exhaust Omar. He was also knowledgeable in the Koran and Muslim law.

This is not all.

He was also an astrologer. We have already said that Omar knew the true value of astrology, but a good dose of information has to be absorbed in order to grasp its rules.

By the way, one of the stories of Omar's astrological feats makes one think that he was familiar with the essentials of meteorology.

The recollection is that of Nizami Aruzi of Samarkand:

"...the Sultan sent to Merv to the great hajji (this is followed by a tremendously long name) to ask the imam Omar to predict the weather and find out, if they go hunting, whether there will he snow and rain on those days."

Khayyam thought for two days, indicated the time, and then "went and put the Sultan on horseback".

From then on, the action in Nizami's story develops like a standard movie. No sooner was the Sultan off, than "black clouds appeared over the land, the wind blew and snow began to fall, and a fog enveloped the earth. There was general laughter, and the Sultan wanted to return, but the hajji imam (Khayyam, that is) told the Sultan not to worry, for there would be no moisture in the course of five days. The Sultan went on his hunting trip, the clouds dispersed, and for five days there was no moisture, and no one saw any clouds."

At the end, the narrator adds that Khayyam, as far as he, the narrator, knows, had no faith whatsoever in astrology. But he had to be able to forecast the weather, because that was one of the standard demands made by sultans upon their wisemen. Consequently, he had some knowledge of meteorology. (I suppose this would be the right place to draw some parallels between oriental sages and 20th century weather bureaus, but I won't.)

So let us add meteorology to the list.

He was, finally, a physician. His biographers have time and again pointed this out.

Besides, Omar busied himself with the theory of music.

And besides all else, he translated from the Arabic into Persian.

Last of all, recall that it was his duty to perform, daily, a host of minor duties for the Shah in the nature of forecasting the weather or interpreting dreams.

Oh yes! We almost forgot, he was also a poet, a brilliant poet.

Now comes the question of when he found time for dalliance with his beauties.

Well, about beauties I am not sure, but about wine he definitely was in the know. Suffice it to recall the highly professional analysis of a variety of wines that Omar gives in the treatise *Nowruz-Nameh.*

However, if all his duties are recalled, one is forced to the conclusion that he had little time indeed to indulge in the worship of Bacchus. Oh, he sinned of course, no question of it. He sinned, but not excessively.

In any case his interests are immeasurably broader than one might think if one focuses only on his quatrains.

The amazing thing, however, is that Omar never says anything about science in his verses. He wrote an autobiography in lyrics, a confession, you might say, yet not a word about what was truly the most important thing in his life.

One might think that such themes were outside the traditions of oriental poetry. Yet wisdom and sages were very often praised. Also, in poetry Omar did not care much for traditions if he handled the almighty merciful Allah in such rough fashion. The only thing in his poetry that can be regarded as referring to science is some skeptical remarks on attempts at learning the meaning of being. Omar Khayyam's world view is by no means so miserable and gloomy.

The only way to tie things together is to presume that Omar was simply showing off to himself by rejecting all and everything and by not finding a single good word even for mathematics. Such coquetry is encountered much more frequently than some are inclined to think. Particularly in the case of poets. There is no reason to be too trusting when it comes to skepticism.

Perhaps more credence can be given to his third cycle of "civic lyrics". Omar seems to have been somewhat of an irritable type, with a rather low opinion of those about him. But try to be calm and good-natured when surrounded by knaves, mountebanks, money-grubbers... if every single day you fear for the future, if it is only your high position at the court that holds in check a pack of thick-skulled scholastics ready to devour you in a moment of weakness, if the position you hold can disappear at any time because of a simple slip of the tongue, or an uncalled-for smile.

Try to be merry and respect those about you if every morning you are not sure how the day will end, if you cannot be like others and if you have to lie every minute, every second and watch others round about you doing the same with evident pleasure. Try all this, and note too that you have no one in whom you can confide, for to share such thoughts is tantamount to a self-imposed exile at best. Try all these things, and if you have the talent of a poet, just see what kind of verses you will produce.

But if, while clearly realizing all these things, you can continue working intensely, remaining a pessimist, a cynic and a drunkard only in poems, but in real life spending your time, energy and nerves in building an observatory, investigating equations of the third degree, writing commentaries on Euclid, studying Aristotle and working with pupils... If you are capable of doing all this, then I will read your verses with pleasure. Especially if they are written in your old age and if loving pupils remain after you.

The year 1092 was the beginning of hard times in the life of Omar Khayyam. In that year, Nizam al-Mulk—his main patron—was killed.

The killing was probably carried out by feudal lords. The murderer was a member of one of the darkest, most fanatical and strange sects in human history: the Ismailians. I recall this for the reason that there is a very curious but obviously unauthentic legend to the effect that Khayyam, Nizam al-Mulk and the founder of the Ismailian Sect Hasan Sabbah all studied at one school and were childhood friends.

In the same year, Malik Shah with whom Omar had been so close also died.

The situation was very bad under the successors, but later he was able to arrange his life. A good deal of money was needed for the observatory, but the subsidies were stopped, so Omar had to make requests here and there. He even had to write a historico-didactic treatise, *Nowruz-Nameh*, where, among a host of anecdotes and tales of eagles, beautiful visages, steeds, and wine is the persistent refrain that

"Malik Shah provided the money for the observatory, and he patronized men of learning".

But, I repeat, things worked out after all. First the son and then the nephew of Nizam al-Mulk became viziers. Probably by force of habit they continued to support Omar.

Meanwhile, the clergy were keeping a keen eye on Khayyam. That he had strayed very far from orthodox Islam was long since evident. Occasionally, the sullen hostility cooled off, but it invariably boiled up anew. Omar had to go in for writing semi-loyal treatises, but that did not help very much.

At times he was intolerant. When he should have kept quiet, he entered into discussions and told sheikhs and imams to their face what he thought of them. Towards old age his temper grew worse, he was sharp-tongued, and still, despite his glory and high-placed patrons, he had to make the pilgrimage to Mecca, the hajj. "And from his hajj he returned to his home town, where morning and evening he visited the place of prayer, hiding his secrets, which will inevitably come to light. He had no equal in astronomy and philosophy; in these fields he was proverbial. Oh, if only he had been given the gift of avoiding inobedience to God."

Thus did the loyal muslim Jamal al-Din ibn al-Qifti write regretfully in his *History of the Sages*.

It is likewise said that towards old age he ceased taking pupils and "grudged writing books".

During the last ten to fifteen years he no longer lived at the court. He somehow displeased the new Sultan and either was asked to resign or was simply dismissed. Perhaps he left of his own accord not wishing to be asked to go. He had no family. The old man was lonely, and the greater part of his gloomiest verses were apparently written during this period.

His pupils were, as before, glad to see him, but he did not seem inclined to receive them.

To all this add the fact that Khayyam was conceited, and with the years his conceit grew; for people of that sort, old age, particularly a luckless old age, is hard to endure.

That he had a very high opinion of himself is acknowledged by his biographers. And his own treatises tell the same story. Even by oriental standards, he would appear to have overdone the self-elevation of his person.

This is how one of his treatises begins: "These are the rays that emanate from the throne of the king of philosophers and the all-inundating pure light of wisdom of the enlightened, skilled, outstanding, elevated, sagacious, great, celestial, glorious, worthy lord of the Proof of Truth and Conviction, the victor of philosophy and faith, the philosopher of both worlds, the lord sage of both Orients Abu'l Fath Omar ibn Ibrahim al-Khayyami..."

Fourteen titles, self-imposed. After that, the beginning of another treatise is a model of modesty: "...the honoured lord, Proof of Truth, philosopher, scholar, seat of faith, king of philosophers of the East and West..."

A rather decent description, title-wise, is given at the beginning of the treatise *Nowruz-Nameh* which was written, as you recall, for the successors of Malik Shah: "...the learned hajji, philosopher of the age, chief of investigators, king of scholars..."

But it is curious to note that all "special"—mathematical and physical—treatises of Khayyam begin in a restrained, dry manner.

Glorification appears in treatises of a general nature. It may be that, to put it into modern lingo, for purposes of publicity he tried to build up his image when the treatise could be read by those that held the strings of power. Naturally, such stratagems added yet another humiliation to the long list of those that Khayyam had to bear. All the more unpleasant to him was this self-advertisement. The last piece of ill luck was that towards the end he experienced real money difficulties.

It is doubtful whether he actually lived in poverty, as some of his modern biographers write. Over the years he held high offices and most likely had some savings. And even to the very end, despite all the attacks of the clergy, he remained the recognized "king of learned men". Also, his numerous pupils could support him if the necessity were real.

So my view is that Omar did not starve and probably lived as prosperously as any small trader. But expenditures had to be cut. At any rate, he complains in a number of quatrains of poverty and of the hard time life was giving him:

If the abodes of bliss be seven or eight
What shall it profit my forlorn estate

Reach me but wine to numb me where I lie
Heart-broken, stretched upon the wheel of Fate.
(Le Gallienne)

The aged Omar was apparently not very happy, the only thing that remained were books. It is said that he died with a book of his beloved Abu Ali ibn Sina in his hand.

One need not think that he was always sighing and grieving, but he was a broken man. Apparently he did not work during the last twenty years of his life, either because he had no strength, or no desire. Life was at an end.

He died in 1128 and even this date was hit upon by accident, thanks to a story related by his pupil Nizami of Samarkand. I give it here in full, for it is far more important for an understanding of Omar the man than all the conjectures of his contemporaries.

Al-Nizami al-Samarqandi relates:

"In 506 (1112/13 AD) the hajji imam Khayyam and the hajji Muzaffar Isfazari were at the court of the Emir Abu Sa'd in the quarter of slave-traders in Balha. We met at a merry meeting. There I heard that the Proof of Truth Omar said: 'My tomb shall be in a spot where the north wind will twice each year scatter flowers upon it'.

"I wondered at the words he spake, but I knew that his were no idle words.

"When in 530 (1135/36 AD) I arrived in Nishapur, several years had already passed since that great man covered his visage with the curtain of dust, the world was without him. He was my teacher. On Friday I went to his grave and took a man with me to show me it. He led me to the graveyard of Hira. I turned to the left and at the foot of the garden wall I saw the grave. Apricot and pear trees of the garden stretched their branches over the wall and sprinkled his grave with so many of their flowers that the ground was completely covered. Then I recalled the words that I had heard him speak in Balha and I weeped, for nowhere in the whole world, from one end to the other, have I seen the equal of him."

We may be quite sure that Nizami was absolutely sincere. It would be hard to believe that thus recalling Omar Khayyam he desired to elevate his reputation in the eyes of the ministers of Islam. But when a man is thus remembered by his pupils, one believes that he was a good man. That apparently was the most important thing. One must believe Nizami, for of all the stories of Khayyam, this one is the story of a friend. Only in this way can we judge the attitude of those who spiritually were close to him.

Very generally, Omar strikingly resembles Galilei in temperament, in views and in many features of his life. It is as if two close relatives lived at different corners of the world separated by an interval of 500 years.

I shall not try to justify this parallel. Anyone with a little pains can do it for himself. As for me, they are as of one kin.

And we finish this story by the same Kipling's wisdom that we started with:

The East is the East. Unlike the West, which is the West...

Chapter 6

The Age of Proofs (*Continued*)

They were many. Very many. No less than a thousand.

One way or another, earlier or later, fortune threw them into company with the fifth postulate and they plunged into the luring labyrinth of theorems.

Not a single one found a way out.

Some were confused from the start, others advanced some distance, but the end was invariably the same.

Some spent their entire lives, others retreated early. Still others went on until nervous breakdown, mysticism, despair overtook them, and yet others philosophically dispatched their sheets of scribbled paper to the waste basket. The end was invariable.

A number followed the mirage and they were happy in the conviction that they had escaped. But the end was still the same.

They had covered the ground of those that came before, without knowing that they were traversing the same false pathways. Hope would flare up at times, and one decisive thrust would seem to have been enough. But again the end was the same.

Dilettantes, professionals, naive mediocrities and brilliant mathematicians; Greeks, Arabs, Persians, Europeans; those that stumbled after the first few steps and those that fought on persistently and inventively—for over two thousand years. They all met the same fate.

The fifth postulate was invincible. It was one of those problems that seemed too hard for the human mind to resolve.

It would appear that mathematicians followed to the letter the motto cut on the grave of Captain Scott:

To strive, to seek, to find and not to yield.

Like the snowy polar wastes, the fifth postulate devoured one after the other.

Most left no traces after them. But there were some who perished nobly, leaving much to remember them by.

In the graveyard of victims of the *fifth* there is one of exceptional honour, Adrien-Marie Legendre.

Legendre was probably the greatest of the mathematicians hypnotized by the fifth postulate. He was engaged in the problem for many long years, attacking the monster from one side and from another. He found evidence and then had to reject it, he proposed proof after proof, passing from confidence in success to despair, still hoping for luck, but at the end he had to admit that no exact solution had been found. The acknowledgement is found in the very title of his summarizing work that he published at the end of his life (1833) *Meditations on Various Methods of Proof of the Theory of Parallel Lines or the Theorem of the Sum of the Angles of a Triangle.*

As often happens in science, this cautious, extensive, and ultimately pessimistic investigation appeared when a solution had already been found and published in the *Vestnik Kazanskovo universiteta* (*The Herald of the Kazan University*)—the first published work of Lobachevsky.

This was not so surprising. A surprising and sad fact was that exactly twenty years later, the Russian Academician Bunyakovsky, who should have been acquainted with the works of Lobachevsky, published a similar study! Note—I wish to stress this once again—note the ridiculous nature of this event. We will come to that again in the next chapter.

In his numerous attempts through the years to prove the fifth postulate, Legendre displayed both persistence and remarkable ingenuity.

Firstly, he proved in elegant fashion a number of theorems of "absolute geometry". Secondly, in proving the fifth postulate via *reductio ad absurdum* he actually found a series of theorems in Lobachevskian geometry. He did not attempt to prove the fifth directly, but rather its equivalent: *the sum of the angles of a triangle is equal to π.*

He started with proving the equivalence.

Even in our home-grown theorem (see p. 43), when the postulate *a perpendicular and inclined line meet* is investigated for equivalence with the fifth, one could already feel how closely tied is the fifth with the theorem of the sum of angles of a triangle.

But at that time we did not give a proof of the equivalence of this theorem and the fifth postulate.

The complete proof of the equivalence of any two assertions contains two parts.

1. One first proves "if assertion A is assumed, then assertion B follows from it".

2. Then one proves the converse: "If assertion B is assumed, then from it follows assertion A".

In our case we have first to prove that if the fifth holds, then the sum of the angles of a triangle is equal to π.

This first part of the proof is a familiar theorem found in all school textbooks of geometry. The second half of the problem was solved by Legendre, and the solution was beautiful. Let us see how he operated. First he proved that

(1) the sum of the angles of a triangle cannot be greater than π.

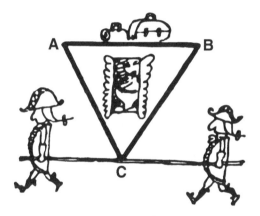

The proof is rigorous. Without any reference to the fifth postulate. He even gives two versions of the proof. Both are correct. The method is the tried and tested *reductio ad absurdum*. It is assumed that there exists a triangle the sum of angles of which is $(\pi + \alpha)$ and it is demonstrated that in this case we invariably arrive at a contradiction. The proofs are rather simple.

I do not repeat them, for lovers of geometry will then have the pleasure of obtaining the result themselves.

Then follow a few auxiliary theorems and he proves a very important proposition:

(2) If the sum of the angles in any one triangle is equal to π, then it is also true for any other triangle.

All proofs are given without invoking the fifth postulate. By means of absolute geometry.

Now everything has been readied for the last theorem of this series:

(3) If the sum of the angles of a triangle is equal to π*, then Euclid's postulate holds.* Once it is proved, the equivalence follows.

[Let us outline this proof. Let I be a straight line and A an external point. Drop a perpendicular from A on I. Let II be a straight line that passes A and forms an acute angle $\alpha < \pi/2$ with the perpendicular AB. We want to prove that this line necessarily crosses the line I. Pick up a point C on I and connect it with

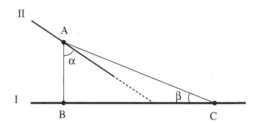

A. We require that the angle β that AC forms with I be less than $\pi/2 - \alpha$. For the points C far enough to the right this condition can always be fulfilled. This *Lemma* can be proved by the means of absolute geometry. It is closely related to the "Aristotle's theorem" (that the distance between two intersecting lines grows indefinitely from the intersection point—see p. 58).

According to our premise, the sum of the angles in the triangle ABC is equal to π. Then $\angle CAB = \pi/2 - \beta > \alpha$. But then the line II is placed between the lines AB and AC and must cross somewhere the side BC of the triangle.—A.S.]

There is only one thing left to obtain:

(4) The sum of the angles of a triangle cannot be less than π. This, nothing more, and the fifth postulate is proved!

Legendre then proceeds to prove it.

The proof he offers is magnificent.

Elegant. Simple. Unexpected.

It contains everything that makes us admire mathematics. With one sole exception.

It is not correct!

Still and all, it deserves our attention.

The method is again that of *reductio ad absurdum*. We have a triangle ABC. This is the our starting point. And the sum of its angles, by hypothesis, is equal to $(\pi - \alpha)$ with some positive α.

Extend the sides of angle A to infinity (we shall need it later).

Now an auxiliary construction. On the side BC construct one more triangle, an exact copy of the first one. It is depicted in the figure—this is $\triangle BCD$. It is so built that $BD = AC$ and $CD = AB$. It is easy to see that

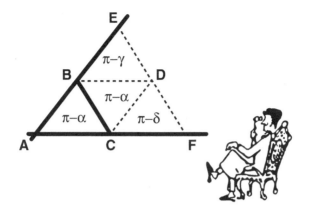

this can always he done. So far the theory of parallel lines does not come into our reasoning in the least.

Now from point D draw a straight line. We impose only one requirement: *the line should intersect both arms of the angle A*. It would seem to be quite obvious that we could find not one but many straight lines that would satisfy that condition.

That is enough. The problem is solved. The fifth postulate is proved. The rest is simply a matter of uncomplicated technique. Take a look at the figure. The sum of the angles of the triangles CDF and BED is necessarily less than π. Indeed, Theorem 1 prohibits it from exceeding π, while Theorem 2 plus the existence of the triangle ABC precludes the possibility of its being equal to π.

How much smaller is quite immaterial to us. More, the only thing we actually need is that the sum of the angles in these triangles should not exceed π. What remains are trifles. Take a look at the large triangle AEF. Find the sum of its angles. This can be done in a rather circuitous way.

We have a total of four small triangles. The sum of all their angles is equal to $2(\pi - \alpha) + (\pi - \gamma) + (\pi - \delta) = 4\pi - 2\alpha - \gamma - \delta$.

Now note that the same sum may be written somewhat differently. Out of the angles of the small triangles, at points C, B and D three angles are arranged that are equal to π in each case. Then there are angles at the vertices A, E and F. But the sum of these angles is precisely the sum of the angles of the triangle AEF.

And so: $\sum_{\text{angles}}(\triangle AEF) + 3\pi = 4\pi - 2\alpha - \gamma - \delta$.

And so the sum of the angles of the triangle AEF is equal to $\pi - 2\alpha - \gamma - \delta$, which is less than $(\pi - 2\alpha)$.

This is followed by a chain reaction. Repeating in literal fashion our construction for the triangle AEF, we build a triangle with the sum of its angles less than $(\pi - 4\alpha)$. Then we construct a triangle with the sum of its angles less than $(\pi - 8\alpha)$. In short, no matter how small α is, we can build a triangle such that the sum of its angles is negative. But this is an obvious absurdity. Our assumption has led us *ad absurdum*. Which completes the proof of the theorem. The sum of the angles of a triangle cannot be less than π. The proof is indeed beautiful. In professional terms, it could be written down in three lines. And only two operations in the auxiliary constructions.

But to presume that through a point inside an angle it is always possible to draw a straight line that meets both sides signifies that in place of the fifth postulate we have introduced its equivalent. And Legendre realized that. But it is such a pity to give up a beautiful solution. So, quite humanly, and somewhat plaintively, he mentions that the angle chosen for $\angle A$ is the one which is less than 60°. Then it is easier to believe his premise. It certainly is easier to believe, but that does not alter matters because it is not possible to prove the assertion without invoking the fifth postulate. So in the end Legendre had to give up his proof.

There is more.

Let $\angle A$ be arbitrarily small. Less than any preassigned number. Less than, for instance, $10^{-10^{10}}$ second of arc. Even in this case it would be impossible to prove Legendre's assumption. If that were possible, the fifth postulate would be proved straightway. It is of course possible to prove Legendre's hypothesis rigorously for points inside angles that are sufficiently close to the vertex. But only for close-lying points, whereas now, in our construction, a contradiction is obtainable only when we go farther and farther away from the vertex.

If the analysis is continued à la Legendre, numerous curious equivalents of the fifth postulate come to light.

Actually, it is thus possible to obtain a large number of theorems of non-Euclidean geometry. Here is a problem for your recreation. In an analysis of Legendre's premise, demonstrate the following: let $\angle C$ be an angle at the vertex of a family of isosceles triangles ACB, $A'CB'$, $A''CB''$ and so forth.

Assuming that *in this family there will always be a triangle with altitude greater than any preassigned number*, we will prove the fifth postulate. A rather unexpected—wouldn't you say?—and quite natural, at first glance, equivalent of the fifth! It emerges rather simply in analysing Legendre's

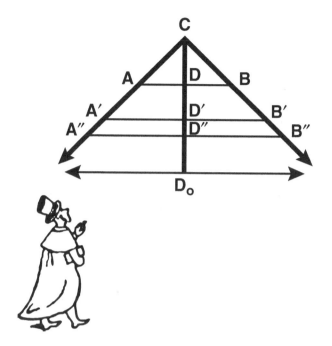

proof. Running ahead of our story, it may be noted that in Lobachevsky's geometry the opposite theorem is correct.

Most of the other scholars did not go so far as Legendre. They got confused at the very beginning.

But there were also more interesting works.

In the year 1889, the Italian geometer Beltrami found a forgotten work of his compatriot, the jesuit Girolamo Saccheri, who as early as 1733 anticipated and surpassed all the results of Legendre.

Up to that time it was believed that namely Legendre had demonstrated that:

(1) Without resorting to the fifth postulate of Euclid, by means of the remaining axioms, it is possible to prove that the sum of the angles of a triangle cannot be greater than two right angles, greater than $180° = \pi$.

(2) If the sum of the angles in at least one triangle is exactly equal to π, then the fifth postulate holds.

Whence the conclusion:

If the fifth postulate is not true, then the sum of the angles in all triangles is less than π.

Legendre wanted to believe that he had refuted this possibility as well, but... well we have already spoken about that.

It turned out that Saccheri had obtained all these results much earlier. What is more, his investigation, his chain of theorems stretches much farther than that of Legendre. True, his starting point was somewhat different. He began with a quadrilateral, not a triangle, just as Omar Khayyam had done a few centuries before.

The construction was as follows:

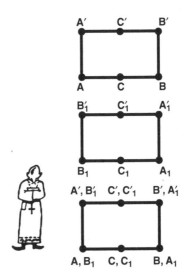

1. Take a line segment AB.

2. Erect perpendiculars at the extreme points A and B and lay off on them segments AA' and BB' of equal length.

3. Connect A' and B' with a straight line. The result is a quadrilateral.

4. Take the midpoints of the bases C and C' and join them with a straight line.

5. Take the "mirror image" of the quadrilateral $AA'BB'$—the quadrilateral $A_1A_1'B_1B_1'$ and superimpose the two so that the side B_1B_1' lies on the side AA'.

It is then easy to prove that angle A' is equal to angle B', and the straight line CC' is perpendicular to both bases. The reader can finish the rigorous proof of this theorem, and he can also obtain this result in a slightly different way—from symmetry considerations.

For angle A' and angle B' there are three possibilities:

(1) they are equal to $90°$ ($= \pi/2$);

(2) they are acute, that is less than 90°;

(3) they are obtuse, that is greater than 90°.

First of all, Saccheri demonstrates that if any of these possibilities are realized in any quadrilateral, then it will be accomplished in all possible quadrilaterals of this type.

He then submits the proofs that:

1. If the "hypothesis of the obtuse angle" holds, then the sum of the angles of any triangle is greater than π.

2. If the "hypothesis of the right angle" holds, then the sum of the angles of any triangle is equal to π.

3. If the "hypothesis of the acute angle" holds, then the sum of the angles of any triangle is less than π.

He then proceeds to prove that the "hypothesis of the right angle" is equivalent to Euclid's postulate.

Consequently, in order to prove the fifth postulate it is necessary to refute the other two hypotheses.

Saccheri handled the "hypothesis of the obtuse angle" with speed and complete rigour.

There remained the "hypothesis of the acute angle." It then transpired that all this was only an introduction, for the real story only now begins.

On over a hundred pages Saccheri investigated the consequences of this truly satanic "hypothesis of the acute angle".

He obtained one theorem after the other, each stranger than the preceding one, but he clearly understood that so far there was no inner contradiction. Then he thought he had it, the proof, the divine spark that would reduce this hypothesis to ashes.

"The hypothesis of the acute angle is absolutely false, for it contradicts the nature of the straight line."

Here it was that the enemy of humankind caught Girolamo Saccheri. He was in error. Crudely.

But no, do not hurry with conclusions. Saccheri was still unsure. He felt something out of order and wrote:

"I could calmly stop at this point, but I do not want to give up the attempt to prove that this adamant hypothesis of the acute angle that I have already uprooted is in contradiction with itself."

The game was thus resumed.

Saccheri again sought proof, but this time in another direction.

He wished to prove that if one accepted the "hypothesis of the acute angle", it would turn out that the "locus of points equidistant from a given straight line is a curved line".

And this is rigorously proved. Note that the conclusion would appear
to be so absurd as to compel one to halt. But Saccheri realized that this
was not yet sufficient.

At this point let us leave Saccheri for a while and recall our honourable
Ghiyath al-Din Abu'l Fath Omar ibn Ibrahim al-Khayyami. It is time to
deliver the goods we promised and relate what he did in attempts to prove
the fifth postulate.

Omar began his proof of the fifth postulate with a critique (as was
usual with all others) of all predecessors. He disproved the efforts of Hero,
Eutocius, al-Khazin, al-Shanni, al-Nayrizi. Also he refuted Abu Ali ibn
al-Haytham (known in Europe as Alhazen) who had taken an extremely
curious and original pathway.

Ibn al-Haytham proceeded from the hypothesis that a line described by
the upper end of a perpendicular of given length is also a straight line if the
lower extremity is moved along the given straight line. (The figure shows
a stick on a roller and a dotted straight line. That is how I attempted to
portray the postulate of al-Haytham.)

Abu Ali ibn al-Haytham himself tried to substantiate this assertion by
reasoning about the properties of motion.

That is precisely what caused certain indignation on the part of Omar
Khayyam. He attacked Abu Ali for introducing motion into geometry. This
is where Omar was not right.

But Abu Ali was likewise in error. Actually, in his proof he utilized an

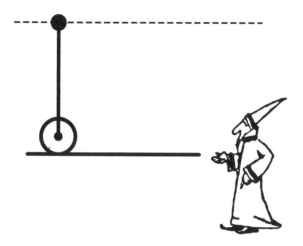

equivalent of the Euclidean postulate already mentioned by us, to wit *the locus of points equidistant from a straight line is also a straight line.* But he had hoped to prove it, not postulate it.

However Khayyam was also punished by Allah for his pridefulness. It was here that he finally fumbled the problem. Unwittingly, he too employed the very same equivalent of the fifth postulate that Abu Ali had. We shall not go into Omar's proof, for it does not stand out among the others. We wish only say that we recalled Khayyam now only to permit ourselves and the reader a tiny lyrical interlude—after all, good mathematicians reason rather well, whether in Greece, in Khorassan or in Italy; no matter that they seek help from Zeus, Allah or Jesus Christ—they strive towards flawless logic and if they err, it is on a very high level. And many of them fully realized that the assertion that "the locus of equidistant points from a straight line is a straight line" had to be proved.

The opposite version may sound strange, but there do not seem to be any inner contradictions in it; the hypothesis will be refuted only when its consequences are reduced to an absurdity.

So Saccheri renewed the struggle.

He analysed the "curve of equal distances" with extreme care, quite rigorously, until—that moment came—the devil led him astray again and he found the proof that it is a straight line. And again he was mistaken. But Saccheri did not see the trap and he was sure the proof was at last accomplished.

That would have seemed to be all, the work was finished, the fifth postulate was proved, and the book could go to the press.

It did. That is, the book appeared a few months after his death (1733) under the sensational title of *Euclides ab omni naevo vindicatus...* (Euclid vindicated *of all flaws, or an essay establishing the very first principles of a universal geometry*).

But the conscience of the scientist was, apparently, still agitated. He wrote in conclusion: "I cannot help but point to the difference here between the above-given refutations of both hypotheses. In the case of the hypothesis of the obtuse angle, the matter is as clear as day... while I have been unable to disprove the hypothesis of the acute angle other than by proving..."

In a word, then, Saccheri was not satisfied. That is clearly felt.

The last trick the devil played with him was vicious indeed. His work remained practically unknown until 1889, at which time it was of purely historical interest.

Actually, Girolamo Saccheri had brilliantly proved several dozen theorems of non-Euclidean geometry, but his preconceptions failed him, constantly making him to believe that the Holy Grail proof of the fifth postulate is around the corner...

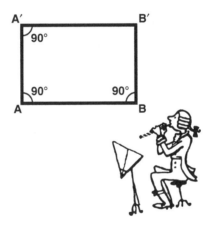

Without knowing of the work of Saccheri, the German mathematician Lambert (1728–1777) went still deeper. He can by rights be considered a direct precursor of non-Euclidean geometry.

Lambert began his analysis by employing a somewhat different quadrilateral. I refer you to the drawing. In it there are three right angles—A, A', and B. Regarding angle B' there can be three hypotheses. That it is acute, right, obtuse.

Lambert rather simply liquidated the "hypothesis of the obtuse angle." We have no time to say how this is done.

But that is not all. Lambert notices that the "hypothesis of the obtuse angle" was justified on a sphere, if one ascribes to circumferences of great circles the role of straight lines. This is an exceedingly interesting and profound observation.

The point is that both Saccheri and Lambert refuted the "hypothesis of the obtuse angle" by rigorously proving that, if it is accepted, the straight lines AA' and BB' are found to intersect in two points.

But this runs counter to a familiar axiom: one and only one straight line can be drawn through two different points.

Incidentally, to reject the "hypothesis of the obtuse angle" it suffices to prove that AA' and BB' intersect in one point. The reader can amuse himself by verifying the latter assertion.

Now on a sphere where two different great circles, like the equator and a meridian, intersect at two points the "hypothesis of the obtuse angle" holds true.

After this slight departure, Lambert returned to the plane. He demonstrated that the "hypothesis of the right angle" is equivalent to Euclid's postulate. Once again it is necessary to verify and refute the "hypothesis of the acute angle".

Lambert began the analysis in the hope of arriving at absurdity and he extended his chain of theorems beyond the point reached by Saccheri.

He proved one of the most remarkable and strange (at first glance) theorems of the geometry of Lobachevsky:

The area of any triangle is proportional to the difference between π and the sum of its angles:

$$S = A(\pi - \Sigma).$$

Here, A is a constant that is the same for all triangles, and Σ is the sum of the angles of a triangle.

From this it immediately follows that the area of any triangle cannot exceed

$$S_{\max} = \pi A.$$

This is the limiting case when the sum of the angles of a triangle is zero. In turn, it then follows immediately that one has only to assume the existence of a triangle of arbitrarily large area and the postulate of Euclid is proved.

It is again clear at once that given the "hypothesis of the acute angle", or, simply, given Lobachevsky's geometry, there are no similar triangles, because there cannot be two incongruent triangles with equal angles.

So the theorem that Lambert proved may be used to propose two new formulations of the fifth postulate.

1. There exists a triangle whose area is greater than any preassigned number.

Or:

2. There exist at least two similar triangles, that is, triangles such that the areas are different and all the angles are correspondingly equal.

(True, as you will recall, this equivalent of the fifth postulate was employed much earlier.)

Both statements are extremely natural and obvious.

There can be no doubt that the elementary consequences of the theorem on areas were clear to Lambert. However, he did not succumb to the sly and delusive charm of the obvious. Quite the contrary, he was enticed by the unmanageable "hypothesis of the acute angle".

"I am even inclined to think that the third hypothesis [the hypothesis of the acute angle] holds true on some kind of imaginary sphere, for there must be some reason, as a result of which on the plane it is so obdurate to refutation, whereas the second hypothesis is so amenable."

That is absolutely correct. Indeed, considering Euclid's geometry to hold on the plane, one can indicate such surfaces that will fully accommodate the plane geometry of Lobachevsky.

These go by the name of pseudospherical surfaces and were discovered by Beltrami. (We shall have occasion later to examine such surfaces, but meanwhile let us see what else Lambert has to say.)

His principal task is to prove that Euclid's geometry holds true on the plane. The remark concerning "imaginary spheres" is a subsidiary conclusion.

And Lambert fully realized—one simply must admire the logic of this man—that he had not proved anything.

"The proofs of Euclid's postulate can be carried so far that what apparently remains is but a trifle. However, a thorough analysis demonstrates that the whole essence of the matter lies in this apparent trifle. It ordinarily contains either the proposition being proved or the postulate equivalent to it."

That is his conclusion, and it is a flawless, precise one.

Without a doubt he disentangled the problem better than any of his predecessors, he carried the analysis farther and enumerated a number of absurd (from the viewpoint of Euclidean intuition) conclusions to which the "hypothesis of the acute angle" led, but he did not find a logical contradiction. And "arguments called forth by love or ill-will" as he classified them are not the arguments of a geometer.

What is more, deep within him Lambert nebulously suspected that perhaps the fifth postulate was, in general, unprovable. He discussed the possible truth of the "hypothesis of the acute angle".

In his enthusiasm for the unwinding chain of his theorems, he unwittingly broke away from his academic style.

"There is something enchanting here that even makes one wish that the third hypothesis be true.

"And still, despite such an advantage, I should like this not to be, for it would involve a whole series of other inconveniences.

"Trigonometric tables would then become infinitely extended, similarity and proportionality of figures would disappear altogether, not a single figure could then be represented other than in absolute magnitude, and astronomy would find matters very difficult."

The words "despite such an advantage" refer to a remarkable conclusion of non-Euclidean geometry: the existence of an absolute unit of length. As we see, Lambert was in possession of this concept too. (We shall come back to the absolute unit of length later on.)

Unfortunately, the work of Lambert was likewise overlooked by mathematicians. To the very end of his days, Lobachevsky knew nothing of it.

It is not clear, however, whether one should regret this or not. If Lobachevsky had known about Lambert's work, it might have saved him a couple of years of work, but it also might have quenched the interest in the problem, for he might have convinced himself that all the initial results had been already achieved.

Be that as it may, he did not know of this work.

There was very little distance to cover for Lambert to become the author of non-Euclidean geometry. Actually, only one thing had to be done.

And that was to state firmly that the "hypothesis of the acute angle" is logically as acceptable as the fifth postulate.

Neither the fifth postulate nor its counterstatement (the "hypothesis of the acute angle" in the terminology of Lambert) follow from the other axioms. They are quite independent. Which one is accomplished in our universe is simply a question of experiment.

One had only to formulate clearly these, one would think, simple thoughts and believe that that is exactly the way things stand, and the rest would have been a simple matter of technique, so to speak.

A mathematician with the endowments of Lambert could have relatively simply proven a few dozen more theorems and could have, with just a little effort, systematized them and thus constructed the entire system of non-Euclidean geometry.

Let us stop here for a moment.

The laws of scientific creativity are hazy. Discoveries are made in a variety of ways; some accidentally, others appear to crown the efforts of years of excruciatingly intense work. Anything is possible. But one law is unalterable. Any ultra-brilliant prevision that is incomprehensible to contemporaries, appears after the passage of fifty years (a hundred, at the most) natural, simple and almost trivial.

In order to appraise a piece of work properly, one has to attempt to shed oneself of the range of knowledge that has since accumulated and mentally picture the epoch under study.

Let us try to imagine ourselves in the shoes of a geometer of the end of the 18th century or the beginning of the 19th century investigating the fifth postulate.

From an early age we are told that the geometry of Euclid is the most perfect creation of the human mind. We are not only taught that but we ourselves, as the years pass, succumb more and more to the enchanting logic of the proofs, sinking deeper and deeper into the cold beauty of the drawings, the lemmas and the theorems, into the moonshiny kingdom of

the logic and intellect.

We live in this closed world, and the only laws governing our thinking processes are the laws of this world. Geometry has long since changed from what it was in ancient times—"the science of the measuring of the land". The problem of its reality, of its practical accomplishments in our world was solved so long ago that today not a person gives any thought to it any more.

Geometry has so long since risen from the sinful earth to the mountain peaks of the ideally abstract...

The very idea that geometry still can and must be verified by experiment, that geometry is actually only one of the divisions of physics can never enter our minds, for at the very beginning of our days at school we learned that geometry has been in man's faithful service for several thousand years.

True, in recent times the entire system of axioms has been undergoing a certain critical review.

True again, the notorious fifth postulate is a shock to our aesthetic feelings. But that is all.

There can be no doubt whatsoever of the truth of the fifth postulate. The only thing we are doubtful about is whether it is a postulate or not. We simply suspect that a theorem has found its way into the axioms.

To suspect the fifth postulate as such, would mean to put the whole of geometry in doubt. And if that were so, then there would be just as many grounds to suspect, say, the axiom that "one and only one straight line can be drawn through two points". Or any other axiom. Then one would have to revise the notion of lines. And the axioms of arithmetic. Then the ideal structure of ancient proportions will turn into a shapeless conglomeration of fragments. That is possible. But that is the work of a barbarian, a vandal, not a mathematician.

There is nothing more perfect in the world than geometry, and here there is only one minute blemish that embarrasses us—the fifth postulate.

As for the other axioms, they are so obvious that no serious problem could ever arise. Slight modifications and more polished formulations? Yes, those are possible. But of no interest, when one comes down to it. That is how we think, that is how mathematicians have thought for the past 25 centuries. To give up this faith is to give up everything we have.

We strive towards beauty and harmony in our Euclidean geometry, and towards an ultimate finish to the edifice it is. Least of all do we contemplate destruction.

And we are firmly convinced that one cannot change a single axiom in Euclid's geometry without arriving at a horrible absurdity, which would explode the whole system.

Just one thought is needed, one phrase, but that thought is such that will change our entire world view.

Chapter 7

Non-Euclidean Geometry.
The Solution

In 1911 the bibliography on non-Euclidean geometry totalled 4,200 works. Today, that number has risen to between 20 and 25 thousand.

Not less than a thousand of these are studies of an historico-biographical nature.

This is only an estimate, of course, but it is based on rather conservative assumptions and one may guess that the actual number of works is substantially greater. We shall take it to be one thousand. Probably at least two hundred books and articles have been devoted exclusively to Lobachevsky.

So why put out one more? That is exactly the question that has plagued the author since he began—even before he began—his work, and it still stood after the book was finished. One consolation of course is that such problems invariably arise no matter what subject for discussion is chosen. Just about in the year 2018 BC, a forgotten ancient Egyptian pessimist and skeptic complained bitterly: "If I could only say something that has not already been said many times before!"

That is slight comfort, all the more so since the flood of writing done in the past four thousand years has nearly drowned humanity, though if we are to believe the classics, a truly great book is written once in a hundred years. But a reasonable person of middle age (the author, by the way) cannot think in such terms. So we are left with the question: why?

Indeed, what can I add to the many many volumes devoted to the history of geometry in general and non-Euclidean in particular and to the general theory of relativity in still more particular?

First of all, we must say that the book is superficial. It is, and could not be otherwise.

Even aside from purely special questions, about two years of hard every-day work would have to be spent in spading up and looking through the more important biographical sources. But that in itself is not yet sufficient. A conscientious biographer has to make a thorough study of all the works of the person in question and investigate painstakingly the response of the scientific colleagues that were acquainted with him and/or his works. He should... there is even more that he should do.

Incidentally, Lobachevsky has such a biographer. It is Academician V. F. Kagan. He wrote a magnificent and profound biography of Lobachevsky. A little too profound, perhaps. It is not very easy to read.

As a dilettante in mathematics (and for a variety of other reasons) I realized that I could not compete in those respects with Kagan. Neither could I compete with many other biographers and investigators of Lobachevsky and of other scientists that have been and will be mentioned in this book.

That brings us straight back to why I wrote this book. It was important for me to know, otherwise these pages would not have been written (maybe that would have been the best option).

But my idea was that nobody had yet written about these heroes as human beings, not as outstanding mathematicians, men of genius, but as normal (or almost normal) people.

And I tried, of course, to write about it so... to convey everything... Look, your author is overwhelmed by strong feelings and cannot continue...

Be that as it may, this is what I want to write about.

And about Work. Real Work of Real Men.

Being a popular-science author, I cannot help resorting to the time-honoured terminology which has worked so well in other books, especially books for the youth.

Strong men advance triumphantly through pages and screens.

Strong men beat up bad scoundrels and conquer the hearts of charming witty sporty girls.

Strong men born in distant villages come to the capital and conquer it as easily as the girls' hearts.

Strong men go from the capital to the countryside and conquer it as... (see above).

Strong men hide their strong feelings under the cover of nonchalant words.

Sometimes strong men booze. This is not typical, but is allowed in difficult moments of life.

Strong men also conquered your author and he is overwhelmed by the desire to write about them.

About real men and not about the characters of, say, Erich Maria Remarque, an author whose novels I do not like so much, even though they are appreciated by the charming girls mentioned on the preceding page. Well, this is with the exception of *All Quiet on the Western Front*, which is a magnificient book.

Excuse me for this improper literary distraction, but I am really kind of fed up by all those dummy characters marching through the pages of many books for two thousand years and more.

So about Work.

And about people who did this work. (Well, I think I've already said that. Never mind, one can repeat.) About people and not about geniuses.

Because, frankly speaking, my feelings toward "geniuses" are mixed too. It is very difficult (though I'm trying) to fight against an established tradition. When people set out to write about Lobachevsky, Gauss, Einstein or Pythagoras (recall the first chapter), in many cases the pages of their books shine with the same blue light of reverence and devotion that emanated from the eyes of Kisa Vorobianinov when he talked to Ostap Bender.[1]

Well, such a reverence towards outstanding scientists could well be understood. In most cases they merit it. And let me make it clear—I am very far from the idea to compare O. Bender with Gauss. Still I believe that such a semi-religious attitude towards scientists humiliates the author, the readers, and the scientists themselves. And if somebody wishes to write their biographies in this style, he may do so, such books will find their readers, but your author will not be one of them.

The very novel idea that I have up my sleeve is that a person should, above all, be a person, a man, a human being. And even such a trifle as a bad temper, a disagreeable disposition and a difficult nature can disperse any kindly feelings stemming from his work.

Starting in this key, I find it hard to decipher my own feelings with respect to János Bolyai.

His gifts were amazing. An inexplicably brilliant talent. His style alone proves that he was a mathematician by the grace of God. It was only later, in the 20th century that works on mathematical logic began to be written in his style. Not a single extra word, ultimately compact, flawless logic, exceptional clarity of reasoning. In the central problem—that of the consistency of non-Euclidean geometry, he advanced farther than Gauss and Lobachevsky. Actually he was very close to the basic idea of proof. He

---

[1] Allusion to *The Twelve Chairs* by Ilf and Petrov.—A.S.

did not find it, but he clearly realized the direction in which it was to be sought.

Here he was ahead of all the rest.

It is quite possible that, for himself, he formulated the ideas of non-Euclidean geometry somewhat earlier than did Lobachevsky. In about 1823.

True, his work was published in 1832, two years later than the first work of Lobachevsky.

But let historians of science debate the priority issues.

For that matter, still earlier a German lawyer Ferdinand Schweikart (at one time professor of law at Kharkov University) had mastered the basic elementary conceptions of non-Euclidean geometry. True, he never published anything, but his nephew, Taurinus, whom he got interested in this problem, put out a booklet.

Though an incomparably weaker mathematician than any in this story, Taurinus came very close to a solution. He developed non-Euclidean geometry in rather some detail, solved a large number of subtle problems, but he did not have a clear notion of the matter. In the end he arrived at the same point that previous investigators of the fifth postulate had—an attempt to prove it and, consequently, the truth of Euclidean geometry.

This is all the more surprising since at the same time he would seem to have an excellent grasp of the consistency of his non-Euclidean constructions, yet...

We have already mentioned the fact that actually only one single idea was needed for the construction of non-Euclidean geometry. Anyone who strived to prove the fifth postulate by *reductio ad absurdum* invariably came to theorems of non-Euclidean geometry. Lobachevsky himself, writing of Legendre, said:

"I find that Legendre time and again took the pathway that I had so luckily chosen."

But it was the basic idea that Legendre (and all other mathematicians for over two thousand years) lacked.

It was first expressed, but was not fully realized, by Lambert; it was stated nebulously by Schweikart and Taurinus; Gauss had been inclined in that direction for a long time without actually mentioning it. It was only Bolyai and Lobachevsky who formulated it clearly.

As to rigour and profundity, the first (and only) work of Bolyai exceeded all others.

Later on, working intensely, Lobachevsky investigated non-Euclidean geometry much more broadly and in far greater detail, but if we compare the first works, the more brilliant is that of Bolyai.

The brilliance of his talent was evident in all things.

He was not only a mathematician of genius, he was an extremely gifted musician. At the age of ten he had already written a number of compositions. Later he became an accomplished violinist.

This does not exhaust the talents that Bolyai possessed. Apparently, he was one of the best fencers of the country. This is no simple matter in any country, but particularly so in Hungary.

Finally, his social views make him closer to us than any of the other personages. He was hostile to all nationalism; an ardent supporter of the Hungarian revolution of 1848, he thought intensely and profoundly on problems of social being. His ideas were akin to those of utopian communism. Towards the end of his life he got attracted by the idea of constructing a mathematical model of an ideal state with the aim of finding a perfect blueprint for universal happiness.

The "theory" was called "The teaching of universal good".

In mathematics he combined the cold reasoning of the fencer with the poetry and inspiration of the musician.

But there is one thing that hopelessly spoils this charming image. Bolyai had one fundamental flaw—his jealous, touchy, egotistical ambition coupled with a very unpleasant temperament. That is what determined the course of his life. In the end it ruined him.

True, I am afraid to be too categorical in such cases, and quite naturally all that has nothing whatsoever to do with any appraisal of his work, but it is important when discussing his attitude towards his fellow men. And Bolyai, I believe, belongs to that category of people who apply essentially different criteria to themselves and to others about them. That is why I do not find him very pleasant. I would like nothing better than to learn that I am wrong.

As to mathematics, his place in mathematical history is clear. Together with Lobachevsky he enjoys full rights as the creator of non-Euclidean geometry.

True, there was yet a third person.

And here it is that we enter upon that arduous pathway of priority litigation, though, in my opinion, such questions merit hardly a hundredth of the attention that they so often claim. But the history of non-Euclidean geometry is of exceptional interest from a purely human stand.

The first to come to the ideas of a non-Euclidean geometry was the "Göttingen genius", the "prince of mathematicians", the "colossus", the "titan", the "first mathematician of the world", no other than Carl Friedrich Gauss (1777–1855). Those were only a few of the numerous titles that he bore during his lifetime, and—there are no two ways about it—they are all deserved.

Gauss was unique among geniuses. As a mathematician he was, without any doubt, far above Bolyai and Lobachevsky. He was simply a scientist of a different category.

So it was Gauss who wrote time and again that the basic ideas of non-Euclidean geometry were clear to him even at the end of the 18th century.

I am positive that he wrote the pure truth. But he did not publish his results either at that time or at any time later. The results Gauss arrived at can only be conjectured from his letters and diaries that were published after his death.

Why did he not publish his investigations? The reason seems to be known, for he himself gives it a number of times.

For example, an excerpt from a letter to the celebrated German mathematician Bessel. It was written after Lobachevsky had published his work. True, Gauss had not yet heard of it.

"Most likely I shall not be able very soon to prepare my extensive investigations into this problem so as to have them published. It may even be that I shall refrain from doing so for I fear the *Geschrei der Böotier* (the cries of the Boeothians) that will rise up when I express my views."

So Carl Friedrich Gauss was afraid of the "cries of the Boeotians".

In this day and age, classicism has to be deciphered. Whether justly so or not, I do not know, but the inhabitants of Boeotia were considered in ancient Greece to be the most dull and thick-skulled of all, and in the age of Gauss and Lobachevsky, the age of classicism, quotations from the classics were much in vogue.

I have always been rather dissatisfied with Gauss' explanation.

It very well may be that he stated one of the reasons, perhaps an important one. But there were undoubtedly others.

Gauss wasn't the kind to hush up a discovery of such exceptional, unparalleled significance for fear of losing his authority. All the more so that he was risking very little, for his authority was so high in the world of mathematicians that if Lobachevsky's memoir had appeared with his signature, all the "Boeothians" would have acclaimed it and applauded non-Euclidean geometry, bowing once more in reverence to the genius of Gauss.

Incidentally, something similar actually took place. Lobachevsky's works attracted attention only after Gauss' death, when Gauss' attitude towards non-Euclidean geometry became known. The new ideas then instantaneously became understood and recognized. If the writing had been that of Gauss there would have been no doubts whatsoever.

It was quite obvious that Gauss did not in the least underestimate his position in the community of mathematicians. I am sure that, like the prince of mathematicians that he was, he could call his vassals to order if there were any unrest, so that the *Geschrei der Böotier* taken all by itself could hardly have frightened Gauss that much.

The crux of the matter lies elsewhere.

Whether Carl Friedrich Gauss was a good or bad man has been under discussion by his biographers for a full century, but one thing is certain: Gauss gave his whole life to mathematics.

To him, mathematics was all. It was just as necessary for him to solve problems as to breathe, eat and drink. It was an instinct. There were no such things as unattractive problems to Gauss. He could spend months on the most routine, monotonous computational job. He could compile tables for weeks on end, and with the greatest of pleasure he would do work that in this enlightened age is handled by technicians, like listing weary columns of figures—for Gauss they were apparently inimitably alluring.

There is not a branch of mathematics that is without certain fundamental contributions made by Gauss. A simple enumeration would cover several pages of text.

He is amazingly like Isaac Newton in temperament, type of character and way of life, and it seems no accident that Newton was his favourite hero. Like Newton, Gauss was extremely ambitious. Yet this was not the ambition that burnt up János Bolyai.

The first requirement was that he himself must appraise his work, he must be positive, and he must be able to say to himself: "Gauss, that is good."

So it was that numerous studies awaited publication for the sole reason that they were not finished, and there was much to do. Gauss eliminated from his life everything that could in any way distract him from his work. Gauss prayed in the temple of a cruel god, he believed with fanatical intensity and, like every fanatic, he was limited.

He was harsh, even cruel, in his attitude to people, though from his own standpoint he was just. But this freezing condescension is quite justifiably perceived as indifference bordering on rudeness. His was a complicated nature, a difficult person, such that can call forth one's admiration, worship, but never love.

Abel, Jacobi, Bolyai are some of the brilliant mathematicians cruelly hurt by Gauss.

But he did not try to offend, and there is no reason why people write that he was a consummate egotist and that he suffered when someone else obtained outstanding results. That is not so. It is definite slander. Gauss always paid full due to the genius of his brethren. But it was not his fault that their results so often simply coincided with what he himself had achieved but had not yet published, and he had not yet published them for there was much still to be done—the work was so often not finished.

Gauss has been reproached, and sorely so, for his review of the work of Abel. Why?

He wrote: "The works of Abel are above my praise, for they are above my own studies."

How can it be that people think that Carl Gauss simply lied? That he never took up similar problems and did not obtain similar results? Or is he supposed to play the part of a noble father? Is it not enough that tens of fundamental theorems which he had proved but, for a variety of reasons, had not published, were published by others so that the fame of discovery had to be divided?

Gauss did not read the papers sent to him for review and did not allow his friends to give him the memoirs of other scientists to read.

He wished to serve his god in such fashion that no one (and above all, he himself) could entertain the slightest suspicion of other people's phrases in his teachings.

His love for mathematics was inseparable from jealousy. This was the love of a man. More, this was the love of a Muslim. And he was cruelly hurt if one of his many "concubines" should so much as smile at anyone else. But he also knew that only the deserved entered his harem, and this consoled him somewhat. He was always ready to be the first to recognize the merits of a rival. But it did not give him joy.

Thus, Gauss lived an even, quiet, monotonous life, while in his brain there continuously rose up and vanished marvellously magnificent, immeasurably more beautiful worlds than that in which he existed.

It is worth repeating that Gauss deserves worship, but it is very hard to love him. In fact, if it were not for Archimedes and Einstein, one might have to accept the fact that a genius of mathematics cannot be other than that.

A hundred or so years ago, I think it was Emerson who said a very curious thing to the effect that each may take what he wants and pay the full price.

For Gauss and Newton the price was extremely high. Einstein and, as far as I can judge, Archimedes, too, received everything that those two had, and got around paying for it.

Another man of the same mould was Nikolai Lobachevsky. Although he was brilliantly gifted, he was a scientist of a different class than this quartet, but to my mind he was much more pleasant than Gauss.

I must repeat that I would believe Gauss to be a superior being, a man of the future or a descendant of a Martian sage, if it were not for Einstein.

One of Gauss' loves was non-Euclidean geometry. What was it that dissatisfied Gauss and why did he not publish his studies? Here we again enter onto the slippery path of psycho-detective analysis, but it is too late to give up.

First of all, the facts.

1. Gauss wrote in his private letters—and there is no reason to doubt that what he wrote was the truth—that the basic ideas of non-Euclidean geometry were clear to him as early as at the end of the 18th century. At that time Lobachevsky had not yet begun studying in the gymnasium (secondary school), and Bolyai had not even been born.

2. The exceptional significance of the problem itself is obvious. It is inconceivable that Gauss could have underestimated it.

3. It is a fact—and we shall come back to this again—that Gauss made several attempts to measure the sum of the angles of a triangle formed by the vertices of three mountain peaks. Consequently, he allowed for the possibility that the geometry of nature might be non-Euclidean.

4. An investigation of Gauss' archives after his death revealed only very

meager sketches, and nothing in the way of a systematic consideration of non-Euclidean geometry.

5. After reading the works of Lobachevsky and Bolyai, Gauss—in both cases—stressed the fact that there was nothing essentially new that he could find for himself.

True, there is a slight complication here. The point is that Lobachevsky gave an incomparably broader view of the possible consequences of non-Euclidean geometry than Bolyai did. In this sense, their works cannot be compared.

For instance, Lobachevsky carried his investigations to a stage that demanded the apparatus of mathematical analysis. One of his works is specially devoted to the application of "imaginary geometry" to the computation of definite integrals.

In the fragments that Gauss left, there is not even a hint that he had reached such problems. Nevertheless, one is led to think that Gauss was perfectly sincere in his letters. If he did not develop non-Euclidean geometry so fully as Lobachevsky, there can be no doubt that he could have very easily if he had wanted to.

He of course foresaw, in principle, all the routes of non-Euclidean geometry into analysis. It is very likely that he could have, without any trouble, developed the scheme of non-Euclidean geometry much more profoundly and fully because his genius and mathematical culture were unparalleled.

This last statement is beyond the shadow of a doubt.

6. Be that as it may, Gauss did not invest his ideas in any kind of finished form and did not publish anything. It is only his letters that show he possessed a great deal.

Let us try to figure out WHY.

We reject Gauss' own explanation, which is about as convincing as the statement of a battleship's commander to the effect that he failed to carry out an important assignment for fear of the adverse reaction of some fishing boats that might be lingering on the horizon.

Well, that may be going too far, but one spectre could have pursued Gauss. To accuse him of incompetence, as Lobachevsky was accused, was out of the question. No one would have dared to. But the suspicion that Gauss might simply have gone mad is a possibility, for one should not underestimate the conservatism of mathematicians (scientists in general, for that matter).

The whole story of non-Euclidean geometry is the best instance of this nature. Even so late as the seventies of the 19th century, when everything

was already clear and the noncontradictoriness of non-Euclidean geometry had been proved, when its ideas had seen brilliant development and were supported and strengthened by the authority of all the greatest mathematicians of the world, there were still professional mathematicians, some in the rank of academicians, that continued to propose all manner of proofs of the fifth postulate and even refused a serious and objective consideration of the geometry of Lobachevsky.

Incidentally, one of the most consistent, implacable opponents of the new ideas was Bunyakovsky, who in 1853 completely ignored the works of Lobachevsky.

However, there is also no need to overemphasize the conservatism of mathematicians. Gauss realized full well that the best scientists, the younger ones first, would grasp and properly appraise the new ideas. Too, he was not the kind to retreat in the face of possible unpleasantness.

Firstly, the most prominent feature of his being was a strict, demanding pride, even arrogance. Secondly, he never betrayed mathematics, for he worshipped it with the frigid passion of a puritan. He would do anything for mathematics, so no spectres would have stopped him.

The next supposition, to the effect that "Gauss did not consider the problem so very significant and for this reason he simply did not have the time to investigate non-Euclidean geometry further" is just as absurd.

That would imply that Gauss was just a mediocre mathematician devoid of much of what is called mathematical culture.

What is more, Gauss' numerous letters that bring in the topic of non-Euclidean geometry, constantly treat it as a problem of the first rank, central to all mathematics.

So why indeed did Gauss not turn his energies and his amazing unparalleled talent to this problem? Why did he remain silent for so many years allowing, in the end, Lobachevsky and Bolyai to outstrip him?

To get things into better perspective, let me give a picture of the whole problem of non-Euclidean geometry.

As you recall, when speaking of axioms and axiomatics, we agreed that only two requirements are imposed on the axioms of any mathematical theory: completeness and independence. The completeness of a system of axioms implies that any conceivable assertion relative to the primary notions can be proved or disproved with their aid.

Axioms permit investigating everything. Let us not go too far into abstract logic, a few concrete examples will better serve our purpose.

Suppose two chess players have studied the game from a textbook that by accident failed to mention a situation in which one of the players cannot make a move without infringing the rules and his king is not under attack. This situation is conveyed by a single chess term: stalemate. Our players would not know what to do. The game could not go on, and they would simply have to introduce another rule, another axiom. In chess this situation represents a draw, in checkers the side that initiates the stalemate wins.

But some new axiom has to be chosen.

Their system of axioms proved to be incomplete, for it did not provide for all possible situations.

One could take football with its Basic Concepts of 11 players, the ball, the referee, goal, field, etc. And again the axioms (rules of the game) have to be formulated so as to be able to judge unambiguously about any possible game situation.

That accounts for the constant arguments and fights that break out in scrub games where the participants do not have a full code of rules—hence the danger of neglecting axiomatics. Though, as a rule, the teams first come to certain modifications of the terms about the game as applied to the local field; setting up a complete system of axiomatics even concerning such a simple game as football is by no means an easy matter. Whence, again, all the tragedies.

Or, to take a final case, the criminal code should in principle provide a complete system of axioms governing all possible situations hazardous to society.

The requirement of completeness would seem to be clear enough now. Would seem! If only things were as simple as I have pictured them here, mathematicians would be in ecstasy.

If I may be allowed a few naive suggestions...

A system of axioms relative to a given group of basic (primitive) notions is complete if for any general proposition A (any theorem) referring to the given primary motions, we can resolve the following question on the basis of these axioms: "Is A true or false?"

Now think over what has just been said. To verify the completeness of the axioms we must do no less than prove or refute *every conceivable theorem*. If that is done, then any mathematical discipline would be exhausted to the end. Exhausted in the same way that the game tic-tac-toe has been.

Our demand is obviously unrealistic.

Even in such a comparatively simple system as checkers, we cannot precisely investigate the basic theorem and answer the question: what result should an ideal game give?

Still less do we know of the situation in chess.

And less still can we provide for and analyse all the theorems of geometry, arithmetic and, in general, any mathematical discipline.

That is the reason why the whole problem of the completeness of a system of axioms must be formulated quite differently.

We cannot here delve too deeply into the depths of higher mathematical logic and so we shall not give in full the problem of the completeness of a system of axioms. Perhaps a beautiful and incomprehensible phrase will suffice: *a system of axioms is complete if any two interpretations of it containing real content are isomorphic.*

The elegant idea of isomorphism was introduced by Hilbert in the beginning of the 20th century. But we will not speak any more about it.

An instance in which a system of axioms was incomplete has already been given, and most likely the reader can think up a few more cases.[2]

The requirement of independence or noncontradictoriness[3] would at first

---

[2] A short remark addressed to pundits: actually, the wonderful theorem proved by Kurt Gödel in 1931 states that *any* complicated enough axiomatic system is incomplete. This refers in particular to arithmetic where there exist equations about which it is impossible to say whether they have or have not integer solutions. V. Smilga mentions this theorem at the end of Chapter 11.—A.S.

[3] In Chapter 3 we wrote that the requirement of independence of axioms is a special case of that of noncontradictoriness (consistency). Some textbooks, however, state that

glance seem to be clearer. Let us phrase the independence requirement rigorously.

Let there be a group of axioms Σ (this letter is ordinarily used to designate a sum).

Let there be some kind of assertion A.

And the opposite assertion is[4] $\bar{A}$.

Then A is independent of the group of axioms Σ if neither A nor $\bar{A}$ contradicts the group of axioms. In other words, both the assertion A and the contrary assertion $\bar{A}$ are compatible with the group of axioms Σ.

All of this is rather elementary logic, though it is probably a bit unusual, and so appears to be complicated. That is why we shall explain everything for the case of the fifth postulate.

We wish to prove that the fifth postulate is independent of all the other axioms of Euclid's geometry (here, the fifth postulate is an example of our assertion A). We express an assertion that is contrary to the fifth postulate (assertion $\bar{A}$). For instance, we state that through a given point al least two parallel straight lines can be drawn to a given straight line. (To simplify matters, we shall write the postulate, which is contrary to the fifth, upside down, like this "Λ ǝʇɐlnʇsoԀ".)

We now have to prove that "Λ ǝʇɐlnʇsoԀ" does not contradict the remaining axioms of geometry. This means that no matter how far and wide we develop the possible consequences, we shall never come to a logical contradiction.

So far so good. Now be careful. How is one to be sure that there will never be any contradiction?

Suppose we have proved twenty noncontradictory theorems. This is no guarantee that a contradiction may not appear in the twenty-first. After proving one hundred, we can expect a failure in the one hundred and first. The same goes for the thousandth. It is quite clear that in this way we will never obtain a rigorous proof of consistency. But we must, for otherwise the problem will remain unsolved. It would seem to be a hopeless task. There do not appear to be any conceivable pathways, other than what we have described. Absolutely hopeless.

Let us stop here again and concentrate for a moment.

---

a system of axioms must satisfy both the requirement of consistency and that of independence. But what is mostly needed in practice is consistency of the axioms. It is even convenient at times to choose some of the axioms not to be independent. Therefore, the requirements of consistency and independence are frequently separated.

[4]The bar on top of the letter is a mathematical symbol to denote the contrary assertion.

In the latter half of the nineteenth century, roughly 20 years after the deaths of Lobachevsky and Gauss, a rigorous proof was given of the noncontradictoriness of non-Euclidean geometry. The proof was unexpected, improbable. We will relate it a bit later.

The point is that neither Lobachevsky nor Gauss even suspected possibilities of this nature. Remember one thing: the very possibility of existence of fundamentally new ideas that would help to prove the noncontradictoriness of non-Euclidean geometry was in those days just as inconceivable as the possibility of determining the chemical composition of a star. Just as inconceivable as overthrowing the mechanics of Newton. Just as unthinkable as a thermonuclear reaction.

There was, at that time, still no clear conception of axiomatics. There was complete chaos in all definitions and axioms of geometry, the disarray that was the legacy of Euclid.

Mathematicians had not yet formulated for themselves practically anything of what has just been written.

It was only the brilliant Bolyai who was groping in the right direction. I am afraid that even Gauss was not fully receptive to his ideas. There was only a semi-intuitive conception about the notions of independence and consistency.

But then—well, then it is clear that it is altogether impossible to prove logically the "independence of the fifth postulate". No matter how long the consistent chain of theorems obtained by means of "Λ ǝʇɐlnʇsoԀ" is, there will always be the possibility that the contradiction is concealed still deeper. There will be a feeling that we simply have not yet reached it.

In despair, of course, one could resort to manipulations that are totally alien to mathematics—experiment. For if it were found—some place in the universe where non-Euclidean geometry works, then the problem of noncontradictoriness would *ipso facto* be resolved.

You recall that Gauss attempted to verify what the sum of the angles of a triangle formed by three mountain vertices is equal to. Quite independently of him, Lobachevsky asked that similar measurements be carried out. Lobachevsky chose a better object. At his request, astronomers at Kazan observatory measured the angles of a triangle whose vertices were three stars. In both cases, the sum of the angles proved equal to π within experimental error.

This result did not refute anything because, even if Euclidean geometry were not accomplished in our world, a deviation from π might be very slight.

As to proof, there was even less of that; nothing in fact.

So what have we? Reasoning in accord with rigorous logic, one thing remained, and that was to conclude that the question was open. And will probably remain so for ever. That, in effect, is what Gauss once said. (In a private letter, naturally.) Here is what he wrote: "I incline more and more to the conviction that the necessity of our geometry cannot be proved rigorously. At any rate, by the human mind for the human mind."

This is open to the following interpretation: "I do not see any conceivable possibility of proving that a postulate contrary to the fifth postulate ("Λ ǝʇɐlnʇsod") does not contradict the other axioms of geometry." And although intuition of course hints to Gauss that the correct answer is *non-Euclidean geometry is just as consistent as Euclidean*, there is no proof.

The problem remains unsolved.

And if that is the way things stand, it is entirely in the spirit of Gauss not to publish his results. He could not risk his reputation and publish a paper of which he was not one hundred per cent positive. He did not possess the idea that would permit cutting the knot and resolving the matter... At this point, factors enter which are not directly connected with pure science.

One after the other, his correspondents (Schweikart, Taurinus, Bolyai) sent him letters which contained a more or less broad hint that it was impossible to prove the fifth postulate and that the contrary postulate did not run counter to the other axioms of Euclid.

As far as Schweikart and Taurinus were concerned, the idea was nebulous and stated in unwieldy fashion. Gauss saw the matter in a clearer light.

Picture Gauss for a moment. It is not so easy to give a direct and honest answer. It is not so easy to present one's ideas to a Schweikart and give up completely the hope, in one's heart of hearts, to resolve that accursed problem, explain the situation, and to advise: develop your arguments as fully as possible, and with the greatest possible care, for the more diversified the corollaries and theorems you get, on the basis of "Λ ǝʇɐɭnʇsoԀ", the more secure will your inner faith be that it is noncontradictory. Examine non-Euclidean trigonometry, try to compute the length of curves in non-Euclidean geometry. Get, for example, an expression for the length of a circumference.

Gauss knew what the length of a circumference would be in non-Euclidean geometry. He gave the formula in one of his letters. But our "ideal Gauss" would of course not write about such a thing to his correspondent.

He would keep silent about his own results, and would outline an extensive programme of research, giving encouragement and support to his young colleague. He would write:

"I myself was attracted to this idea, but, alas, no matter how far you develop your theorems, the question—ultimately—of the noncontradictoriness of non-Euclidean geometry is a question of faith. It is impossible to obtain a rigorous proof. One can only rely on one's intuition.

"The probability of error will always remain. You are young. Your name is not canonized, you can afford to write silly things. I insistently advise you to devote all your energies to this problem. Awaiting further letters..."

Aren't we expecting too much of Gauss?

A lot, but not too much.

Science knows of such people and such cases.

The phrase "you are young enough to write silly things" was actually said once—by a remarkable man, teacher and physicist, Ehrenfest, to two young men, Uhlenbeck and Goudsmit, when they had wanted to withdraw publication of a paper they had sent to a journal. Later, it turned out to be their chief contribution to science. Incidentally, they got the most fundamental reasoning from Einstein, who gave it unselfishly, caring not a whit about his own priority in the matter.

But Gauss was not the ideal of scientific disinterestedness. True—this we must say—he never permitted himself any improper actions either. He was always scrupulously honest. Well, nearly always so.

Because in the case of non-Euclidean geometry he never explained himself fully and never gave the true reason for not wanting to publish his work.

In all his letters he childishly insisted on his fear of the "Geschrei der Böotier". These Boeotians, like lifesavers, turn up in almost every letter dealing with non-Euclidean geometry.

I admit even that Gauss himself finally began to believe his pet excuse. But does that change anything? Nothing at all. One of the most subtle, convincing and widespread types of lie is that which you yourself have come to believe in.

Faith is needed; that precisely is what convinces others.

Non-Euclidean geometry is likewise a product of faith.

Bolyai and Lobachevsky believed. Strictly speaking, in the heart of the matter they reasoned as poets reason, and not as worshippers of rigorous logic. "This is correct for it is beautiful" would seem to be their chief argument.

At this point, your author cannot resist the temptation to philosophize a little.

I've just written "reasoned as poets reason". But it would have been better and more correct to have said "like mathematicians", or more precisely still, "like people endowed with creative thought".

The nature of the creative process is unitary in its basic and decisive features. Mathematicians, physicists, poets, artists, engineers, musicians differ among themselves to a far smaller degree than is generally thought.

Incidentally, the ancient Greeks reasoned more exactly in this matter, for they hardly at all distinguished the nature of the different types of creativity. They may have overstepped the limits when they claimed that a musician needed professional training in philosophy and mathematics. But this exaggeration grew up on a basis that was sounder than that of the opposite view.

True, it must be noted that a sharp demarcation between the exact sciences and the arts cannot unconditionally be considered the stand of our century. It is simply a very common view, one held mostly by those who have no contacts with any area of creativity.

Quite naturally, to explain to such people the nature of the creative process is an extremely difficult task, the difficulty progressively increasing with the official standing of the person with whom you are arguing. It is just as hard as to explain to a lover of ballet that a magnificent footballer is no less worthy of admiration than a brilliant prima balerina. And if one adds that, essentially, the artistry of our centre forward and of the prima is of a similar nature, unitary in its very essence, in its objectives and results, the intellectual ballet lover will most likely walk out of the conversation. And if you say the same to a football fan, you would get the answer: "Football is not ballet", plus some unprintable comments on your mental status.

All the more reason for wiping out this dismal, settled narrow-mindedness—it is very widespread.

Let us return to geometry. One of the chief criteria of any type of art is, as we well know, beauty. The search for beauty permeated the whole life story of the fifth postulate, from Euclid to Lobachevsky. The ugliness of Euclid's postulate predetermined the futile two-thousand-year attempts to prove it.

And the elegance of the constructions of non-Euclidean geometry won the heart of Lambert, almost convinced Gauss and compelled Bolyai and Lobachevsky to declare: this is so beautiful that it has as much right to live as the geometry of Euclid.

By rights, Bolyai occupies first place when it comes to faith and enthusiasm. His work entitled modestly

Appendix containing the science of space that is absolutely true and independent of the truth or falsity of the XIth axiom of Euclid, which, a priori, can never be proved...

is written in the most unequivocal terms.

Incidentally, this floriated title provoked an unexpected by János train of events.

The work was published as an appendix to a textbook of geometry written by his father, Farkas Bolyai. As was habitual in those days, the book was written in classical Latin, the language of scholars and philosophers. And of the long title, only the word *Appendix* remained when the work was quoted. That is now the title we know it by.

It is curious and symbolical that at the cradle of non-Euclidean geometry there clashed three human and scholarly temperaments, and three scientific modes of thought.

Opposite stands were taken by Gauss and Bolyai.

Carl Friedrich Gauss. Gauss was a cautious realist. He was undoubtedly the most logical of the three. The most academic. To him the problem had not been solved to the end, and he could not allow himself the luxury of following his intuition, to have faith without proof; that he could not do. He had a clear conception of the matter and, given the desire, he would probably surpass Bolyai and Lobachevsky. He knew it but he did not believe in it enough. And he lost out.

It matters little what historians will write later. It matters hardly at all that in all his letters he insistently repeated "I have known this for forty years already". Alone, by himself, Gauss admitted that he had been left behind. What is more, he was honest enough and severely strict in his attitude towards himself to admit this unconditionally. He had lost the game.

János Bolyai. Bolyai was a romanticist, struck by beauty and elegance, carried away by his own talent, enthusiastic beyond measure. "This was done by János Bolyai!" His faith was rewarded. It was the prime mover of his life. In his first work he grasped the problem more profoundly than anyone hitherto. Incidentally, he never made any more headway. Possibly because for him everything was solved. Subconsciously, perhaps, but solved.

He achieved his goal, he was a genius. That much he had proved.

Nikolai Ivanovich Lobachevsky. In our story he is close to the ideal scholar. Combine in equal measure the scientific enthusiasm of Bolyai and the skeptical cautiousness of Gauss, and to this add a persistence bordering on stubbornness, an almost instinctive inner conviction of the irreproachable truth of his ideas... Also make the demand that this scientific integrity not waver during twenty years of a complete lack of understanding by his colleagues, a lack of comprehension that at times took the form of open mockery, and you will have an approximate picture of Lobachevsky creating the foundations of non-Euclidean geometry.

He believed and he verified his beliefs.

It is quite fair that the non-Euclidean geometry of which we are speaking is always called Lobachevsky's geometry.

We shall return to Gauss and Lobachevsky, but first let us take up Bolyai.

I have already said that as a person Bolyai was not very pleasant. We leave aside an obvious fact that he was a brilliantly gifted mathematician. This he magnificently demonstrated and there is no maybe about it. But apparently Bolyai the man was a difficult case.

He was of the species of "geniuses". Every school has two or three "Newtons"—talented youngsters, sharp-witted, far advanced, towering over all the other children, and with sparkling lightning-fast minds. All too often, recognition of this intellectual superiority spoils them and brings them to a kind of Nietzscheanism. They are temperamental, intolerant, egotistical, trust only themselves, and regard all others as the gray mass, the rabble whose job it is to hoist their hero onto a pedestal.

Without a doubt, there are times when they are kind, responsive and charming, but subconsciously (and, later, even consciously) their philosophy is that of "leaders" and "masses". This kind of development of gifted children is saddening, but it is natural perhaps because the education of an inner culture is a much more lengthy, complicated and subtle process than even the flowering of a talent. And the conflict between a talent and the culture is the more acute and lacking in compromise, the sooner the superiority of the child becomes evident. If I may permit myself some philosophizing, one is inclined to think that most of the tribulations of mankind are associated with self-complacency, which, alas, is a practically inalienable feature in most people. And if a man is gifted and also ambitious, self-complacency may develop to arrogance and life becomes arduous either to himself or to those about him, or to both.

The latter applies to Bolyai. His talent appeared early and in a diversified fashion. He was an extreme representative of the kind of temperament that is usually described as "artistic", "poetical". Elegant, impulsive, scintillating.

The supreme proof of his mathematical talent and intuition is that by the age of 21 to 23 he had already mastered the fundamentals of non-Euclidean geometry and, what is most important, was apparently fully convinced of the truth of his ideas. When Farkas, his father and a prominent Hungarian mathematician, and, incidentally, a school-day friend of Gauss, learned that his eighteen-year-old son was captivated by the theory of parallel lines, he wrote in desperation to his son, pathetically imploring him to give up that mad venture.

The letter is written in such high-flowing style as to irritate the modern reader and cause him to doubt the sincerity of the writer. Of particular interest, in my opinion, is that it gives an excellent picture of the relations that obtained in the Bolyai family.

"I implore you not to attempt to surmount the theory of parallel lines; you will waste all your time on it and still not prove the proposition. Do not try to overcome the theory of parallel lines either by the method you speak of or by any other method. I have studied all avenues to their ends and have not encountered a single idea that I have not developed. I have passed through the whole hopeless darkness of that night and have buried in it every beacon, every pleasure of life. For God's sake, I implore you, leave this matter alone, fear it no less than sensual passions, for it is capable of depriving you of all your time, your health, peace of mind, the entire happiness of your life. This hopeless darkness will never be clarified here

on earth and the miserable human race will never wield anything perfect
even in geometry. This is a great and eternal wound in my soul...."

Indeed, Farkas—in his youth—did study the theory of parallels and
even sent Gauss some proofs of the fifth postulate. There can be no doubt
that the father was sincerely upset about János. Strange to say, starting
from incorrect premises be correctly foresaw the final result: the theory of
parallels was indeed destined to be the curse of János Bolyai's life, though
for quite different reasons than his father supposed.

When there is a devilish obsession, so there must be an evil spirit. Gauss
was the evil genius for János Bolyai from early childhood almost to the end
of his days; though—speaking objectively—Gauss was hardly to blame in
any respect.

It all started when the father began to harbour the ambitious dream
of sending his talented son to Göttingen to complete his mathematical
education under the guidance of Gauss. Farkas wrote to his kind old friend
asking him to receive his son. He was naturally prepared to pay all the
expenses.

The answer was silence.

Gauss may have had a variety of reasons, some very weighty, to refuse,
and he can only be reproached for a lack of tact with regard to Farkas.
Admittedly it is very difficult to judge. Farkas' letter was somewhat imper-
tinent. Some of the questions were reasonable enough, but one can easily
understand Gauss too. "Is your wife an exception to the whole female sex?
Does her mood change like a weather vane?" The point was that János
would have to live there in her house, and so Farkas wanted to know how
János would get along. Quite naturally, Gauss must have winced at such
sweet ingenuousness.

However, I am not interested here either in Gauss or in Farkas Bolyai.
As far as one can gather, Gauss would not have taken an unknown boy even
if the letter had been written with the diplomatic elegance of a Taleyrand.
That is his right. All this gossip is of interest only in that it once again
demonstrates how little and how poorly adults understand children. Both
grownups in this story are to blame.

Imagine a high-strung fourteen-year-old boy in whom his effusive father
had undoubtedly instilled great hopes. The boy did not know much about
the relationship between his father and Gauss. He had no idea of what
could offend Gauss and why. The only thing he knew—and you can be
sure the father spoke of it several times a day—was that as students, the
father and the great Gauss had been the best of friends and that they had

even solemnly sworn to eternal friendship.

So, naturally, the father was convinced that Carl will answer the very next day. Fourteen-year-olds believe their fathers. Especially if your father is also your teacher and is a talented, versatile interesting person. One must add that Farkas was a profoundly gifted mathematician. In his textbook of geometry, he clearly formulated for the first time the demand that axioms be independent. He doubtlessly deserves full credit for János' deep understanding of problems of axiomatics at the age of twenty.

The boy could not help respecting his father.

He was confident and he waited.

He—a boy from a backwoods province of Europe—already saw himself a student of the great Gauss, and perhaps, later, his associate in science. For months on end he waited expectantly for the postman checking the days, adding on when too many went by, waiting for Gauss' answer, thinking up fresh reasons for delays, again waiting and hoping; still hoping when his father took him to Vienna to a military engineering academy, for it had become clear that Gauss would not reply. Gauss simply did not want to. Yet there lingered the hope that, perhaps, an unknown messenger would come riding at their heels with the long overdue letter. No letter ever came.

I must say that though I have absolutely no facts to go by and I do not know how all this affected János, I can easily see how a month or two of waiting like that could totally derange the nervous system of a high-strung fourteen-year-old boy. Particularly if the boy was gifted, excitable, deeply sensitive.

But let us not be overstrict in judging Gauss. He might easily have been offended. And to worry about the nerves of some unknown youngster, as we do so now.... Let us not ask for too much.

The years as a student and especially, the years of military service in outlying garrisons of Hungary were years of dismal aloneness for János Bolyai. True, he had a couple of friends at the academy, brought together by their love for mathematics. Nothing more. Afterwards, there was no one.

I do not think that provincial officers offered him appropriate company. Apparently, he not only failed to conceal his haughty disdain for the whole crowd, he went to lengths to stress it. The result was constant quarrelling and duels. He saved himself by his skill at swordplay. He most likely was right in his attitude to his "colleagues". Yet during all those years he could have found a few decent fellows, even though somewhat lacking in education and intelligence. Of that there can be no doubt. He obviously presumed that they would be of no use to him. He was mistaken. But he was not mistaken when it came to the theory of parallels. Before his resignation (again the result of some kind of scandal) he had written up his investigations in the form of the celebrated "Appendix".

The work is written in extremely compact form and makes difficult reading.

That in general was the poor luck of non-Euclidean geometry. Lobachevsky's papers are also written hazily, and if one judges from the standpoint of a scientific editor, they are simply no good. Numerous essentially simple matters are tangled up beyond measure. For such things mathematicians have the aphorism: "the reputation of a mathematician is determined by the number of unwieldy proofs that he has concocted".

The point seems to be that path-makers, as a rule, do not find the simplest and most elegant pathway. They slash through the trees cutting a road because they have to advance. Those who come later bring elegance, beauty and polish. There are exceptions, but they are truly exceptional.

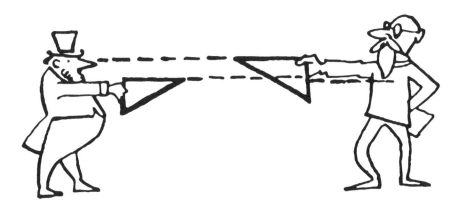

Be all that as it may, Farkas Bolyai did not understand the work of his son at all. Since it was thought to be published as an appendix to the geometry textbook which Farkas had written, the conflict reached its apex.

It is here, after fifteen years, that Farkas again wrote to Gauss asking him to act as judge. (This was in 1832.) "My son respects your opinion more than that of the whole of Europe", he wrote.

This time Gauss replied. True, a month later. But he read János' paper carefully and favourably. Whatever else may be said of him, he valued talent. And in others too. Almost the next day he wrote to a friend of his: "A few days ago I received from Hungary a small paper on non-Euclidean geometry; in it I found all my own results carried out with marked elegance."

Well, such were the facts. Almost the actual facts. We have no right to blame him. Almost no right.

Then the father and son received his reply. The usual introductory remarks and generalities, and then:

"Now a bit about the work of your son. If I begin by saying that I ought not to praise his work, you will of course be amazed for a moment, but I cannot do otherwise, for to praise it would mean to praise myself. The entire contents of the composition, the path that your son has taken, and the results that he has obtained almost completely coincide with my own attainments, which in part are already 35 years old. I am indeed extremely amazed. My intention, regarding my own work, which incidentally has but slightly been put to paper up to the present time, has been not to publish anything during my lifetime. Most people do not take the proper view of the problems discussed here. I have found only a few people that evince special interest in what I had to tell them on this subject. In order to be in a state to master this, one has to feel with great vitality what is, properly speaking, lacking here. Now this is not clear at all to most people. However, my intention has been to write all this down, in good time, and in such form that these ideas should not perish with me. Thus, I am exceedingly surprised that this job has been taken from me, and I am pleased in the extreme that it is precisely the son of my old friend who has anticipated me in this remarkable fashion."

To say that János Bolyai was distressed is to say nothing. He was enraged, obliterated, crushed. He was convinced that Gauss' whole letter was one lie from the first word to the last. A lie, the sole purpose of which was to arrogate János' brilliant idea.

This second blow from Gauss was heavier than the first. He, János Bolyai, had reached what he had sought. He had become a mathematician. He had grasped what hundreds of the greatest geometers had failed to understand over the past two thousand and more years. He alone in the whole universe knows the answer (but he did not know that somewhere on the boundary line between Europe and Asia a certain Lobachevsky had already published a paper with the same conclusions). And this arrogant old man wanted, so he thought, to snatch up the work of his whole life, his glory, and to bury his genius.

Yet one should not reproach Gauss overstrictly. He wrote the truth.

Almost the truth. He dissembled only when he tried to explain why he had refrained from writing up his results and publishing them. Too, there can be no doubt that Gauss sinned both before mathematics and before Bolyai, and also before himself in that he did not express any opinion in print concerning the work of János, for in this he would not risk his good name, he risked nothing. He was led in this, either consciously or subconsciously, by the logic of ambition. Though János' rage was unjustified in many ways, he keenly perceived that Gauss was manoeuvring in some way, that there was an unpleasant, false note in all his reasoning.

We have some notes that convey János' reaction to this event, and we can agree unconditionally to the whole text. The words about science and the ethics of the scientist are good and proper. Here, his accusations levelled against Gauss are fair in full measure.

"In my opinion and, I'm convinced, in the opinion of any unprejudiced person, all the arguments given by Gauss to explain why during his lifetime he does not want to publish any of his own works on the subject at hand are completely impotent and trivial, for in science, as in everyday life, the task is precisely that of illuminating sufficiently necessary and generally useful things, particularly those which are not quite clear yet, and of awakening in every possible way the still deficient or even slumbering awareness of the truth.... To the general detriment and misfortune of all, an understanding of mathematics is unfortunately the lot of only a few; and on those grounds and for those reasons, Gauss could have kept to himself a still more substantial portion of his splendid studies... An extremely unpleasant impression is created by the fact that Gauss, instead of expressing, relative to the *Appendix* and the whole *Tentamen*, a direct and honest acknowlegment of their high value so as to think of means to open the way wide for a good undertaking—in place of all this, Gauss strives to avoid the direct pathway and hastens to pour forth pious wishes and regrets concerning the insufficient education of people. That, of course, is not what life is ..."

But alone by himself, Bolyai did not reason so broadly. He suffered, aspiring to fame and recognition. Recognition is what he wanted. He wanted the whole world to see that he, János Bolyai, was a "geometer of genius of the first rank" (that was how Gauss described him in one of his letters, but not in a letter to Bolyai and not on the pages of any journal).

What Gauss' letter resulted in was a nervous breakdown for János Bolyai. He even suspected his own father of betrayal.

I can't say that I am particularly delighted with the reaction of János. One can of course understand him, but one finds it hard to agree with and justify him. If he had paid heed to his own words about science, his conduct and future life would have been different. Bolyai was then no longer a boy, he was thirty years old and he could have taken all these things like a man. He could have. But, too, let us not judge Bolyai harshly. He was not yet crushed. He continued working on the same problem that, a few thousand kilometers away from him, Lobachevsky was engaged in. He was constructing the whole of geometry on a new foundation.

But the intensity of his work had dropped. He still took a lively interest in a great variety of problems. Together with his father he dreamed of constructing a universal language; he tried his hand in other branches of mathematics; he tried other things too, but none of these was really normal serious work—only a morbid desire to do something out of the ordinary, to prove to the world that he was indeed a genius.

Meanwhile his relations with his father had become extremely bad. Obviously, Bolyai the son was not capable of being a co-author. True, Bolyai the father was far from a paragon of wisdom and good will. Mutual scientific jealousy and an assortment of muddled affairs culminated most unusually. On one fine day, the reverent son challenged his father to a duel. Later still, János became a nerve-patient in the full clinical sense of the word.

The decisive blow was dealt once again by that accursed Gauss.

In 1841, on Gauss' suggestion, Farkas Bolyai ordered a booklet by Lobachevsky published in German and entitled *Geometric Investigations in the Theory of Parallels*. Recalling János in connection with Lobachevsky's work, Gauss was possibly making amends for his long disregard and, undoubtedly, was guided by the very best of intentions.

But János' morbid mind viewed all this as a Machiavellian intrigue on the part of Gauss. He was convinced that this mythical Russian pseudonym simply concealed one of Gauss' myrmidons, if not Gauss himself.

János Bolyai subjected to analysis every comma of this tiny piece of writing; he did it thoroughly, punctiliously; with an ill will he subjected it to a thorough cavilling criticism.

He was scientist enough to appreciate the work, but he was glad of every fault and regarded the author as his personal enemy.

He was then thirty-nine. In his prime.

He was destined to live another twenty years. But he was already broken and crushed. His illness was a form of nervous disease. He was haunted by the theory of parallels. Those twenty years were awful years both for him and those close to him. The rupture with his father was complete. Their only correspondence was on scientific topics. They corresponded, though they lived in the same town. And it was mathematics that finally set them at loggerheads. For the last time, the retired captain, János Bolyai, came to life in 1848 during the Hungarian revolution with which János sympathised completely. But he was ill. That was one thing. The other was that he did not wish to be a rank-and-file participant—only a leader. By the way, one can believe that he could have been a splendid military leader. But he was unknown. And so he remained at home. The defeat of the revolution was yet another blow. His illnesses tortured him, and he no longer worked.

During the remaining years of his life he did practically nothing, only busying himself with utopian ideas. It is remarkable that the brilliance of his talent continued to shine even in this production of his affected brain. One of the last of Bolyai's passions was the construction of an ideal mathematical theory of the state and his hope, in this way, to lead humanity to universal good. Of course, he was unable to do anything of real value here, but the idea itself was very close to modern conceptions of certain cyberneticians.

The end was close.

He was morose, suspecting, and though he loved humanity at large, he could hardly get along with his closest friends. He left his wife; his children ceased to interest him. Once more, for the last time, he quarrelled with his old and dying father. But now, at 54 he himself was an old man.

He would have been happier if he had died earlier.

He was a brilliant mathematician, no question of it. But what he valued above all else was not science, but himself in science. And cruel though it may sound, I am afraid that he himself was the maker of his fate.

Chapter 8

Nikolai Ivanovich Lobachevsky

By the start of this century, Nikolai Lobachevsky had already been canonized.

He was the pride of Russian science. He was the greatest talent in the history of mathematics, despised by his compatriots who did not understand him. He was the victim of a bigoted, beaurocratic academic clique. He suffered the whole of his life and died nearly in poverty, an unrecognized genius.

Such, in brief, were the broad outlines of the cheap melodrama that so often comes to life on the pages of popular journals and books. The remarkable thing is that all this is in some sense true, though grossly exaggerated.

One thing is unquestionably true, and that is that Lobachevsky is indeed the pride of Russian science. The reader would do well to read V. F. Kagan's marvellously detailed and profound biography of Lobachevsky. I highly recommend it. For our purpose here, a few highlights of his life will suffice.

Kazan. An outlying, but not too far outlying district of the Russian Empire. The year 1856.

On February 12, after a protracted illness, Nikolai Ivanovich Lobachevsky died. Shortly before, due to ill health, he had left his post of trustee of the Kazan School District. For many years he had been Rector of the Imperial Kazan University, honoured professor of pure mathematics, Corresponding Member of the Göttingen Royal Society, honorary member of the Moscow Imperial University and also of many scientific societies. Being a hereditary nobleman and an Active State Councillor,[1] he was bearer of the orders of St. Stanislav, Third and First Degrees; St. Anne, Second

---

[1] That was the civil rank of the 4th class according to the Russian *Table of Ranks*, equal to the rank of Major-General in the Army and Rear Admiral in the Navy.—A.S.

Degree; St. Anne, First Degree adorned with the Emperor's Crown, and the Order of Equal-to-the-Apostles St. Prince Vladimir, Fourth Degree and Third Degree; repeatedly noted for outstandingly zealous service and especial efforts by the Supreme Grace of the Monarch.

The funeral was solemn and beautiful, for he was loved and revered in the city. The speaker said: "His noble life was a living chronicle of the university, of its hopes and strivings, its growth and development."

The Kazan Gubernia Vedomosti, the local newspaper, gave a brief obituary in moderately solemn style as befits such an event.

One speaks well of the departed or one does not speak at all. And after a short enumeration of his merits: "His work and attainments in the field of science, which are now in the chronicles of the scientific world, will without doubt find a worthy judge. We, for our part, are happy to be able to adorn in these few lines the memory of the deceased."

One speaks well or one does not speak at all.

The writer of the obituary most likely was sincere in congratulating himself for the clever rhetorical figure that saved him from an embarrassment. Everyone in Kazan knew that professional people and authoritative critics regarded the works of Lobachevsky as the product of sick mind. For many years, the reply to an enthusiastic students query "Is it not true that our rector is the first mathematician of Russia?" was professorial silence. An awkward, sullenly embarrassed silence in the case of well-wishers, and a sarcastic silence in that of his opponents.

The late professor was, undoubtedly, among the most worthy citizens of the city of Kazan. He was an excellent administrator. He was paternally strict with the students, friendly with his colleagues, a skilled diplomat with the mighty of the world, a highly esteemed teacher, an extremely erudite mathematician, zealous in his running of the university, its pillar and its pride.

Yet there was one blemish. His ridiculous works, the monstrous belief, over so many years, in those mad ideas of his. One could only tactfully remain silent.

For those who knew, the obituary notice included a tiny deeply concealed ambiguity "will without doubt find a worthy judge".

Irrespectively of the will of the writer, one could discern a hint here. Not a well-wishing hint either, for surely everyone knew the true worth of the deceased's works, which had been justly evaluated by the best Russian academicians, and this evaluation was not jubilant.

Unfortunately, we must admit that this unpleasant rock still projected above the surface. Neither was Bulich, who pronounced the brief funeral oration, able to circumvent it. A professor of philology, he very properly confined himself to a single smooth phrase, "... It is not for us here to speak of his original scientific studies in mathematics that brought him renown and glory..."

All the rest was said cordially, simply and well, and the sincere, well-wishing educated speaker concluded in elevated moving words with even a touch of poetical fervour.

But again all this was equivocal, even unpleasantly ambiguous, for his renown in the world of science was of a joking kind. God save us from such glory.

Nikolai Ivanovich had indeed stumped his friends, for they had to say something (after all, he was a mathematician, not just some official), but what?

One can add that Bulich was unlucky with his speech by a different reason. In some marvellously strange way, by some superior sense, the archpriest perceived a crime in the funeral oration, a crime against censor-ship, against morality—atheism to put it simply.

How he perceived it is not clear. He was probably indignant that no word was said of divine affairs, not a word about God was mentioned.

And so of course Bulich was duly reported to the authorities in very high spheres. He wrote to friends imploring them for help and assuring them that he had not said anything unlawful "except the truth regarding the deceased, except respect for thinking and science that are so natural today, and except unavoidable rhetorical figures".

Luckily, there were benefactors in St. Petersburg and the affair was hushed up.

That was in the Winter of 1856, when, as Bulich put it, Kazan accom-panied their pride, their great citizen "to the deserted road to eternity".

Only a year and some later did one of Lobachevsky's pupils, A. F. Popov, write an obituary and solve this difficult problem in the best fashion. Again a single sentence to cover a lifetime of work: "The lectures Lobachevsky delivered for a select audience in which he developed his new foundations of geometry must in all truth be termed profound."

Actually nothing said, yet no adverse innuendoes either.

We touched upon now the main tragedy of Lobachevsky's life. One need not talk more. The atmosphere of his funeral and the obituaries that followed it explains more than does any collection of exclamation marks and tragic phrases.

Let us forget for a moment that he was a brilliant mathematician. Let us appraise the initial (and terminal) conditions with the undemanding yardstick of the philistine.

Nikolai Lobachevsky was born on November 20, 1792, in a rather impoverished family of a Registrar, I. M. Lobachevsky. This was the last 14th class in the table of ranks of the Russian Empire, it was equivalent to that of second lieutenant in the Army. Wrote one of his contemporaries, with the modish romantic melancholy of those times, "Poverty and want hovered over the cradle of Lobachevsky."

There were three boys in the family, and when, in 1797, the breadwinner Ivan Maksimovich died, the still young twenty-four-year-old hardly literate mother was on the brink of a catastrophe.

By what means and ways she was able to send all three to the Kazan Gymnasium, and at government expense to boot, what all this costed her really, what tears and what devious dealings, we shall never know.

All that remains is an application written for her either by a kind soul or for a glass of spirits by some heavy-drinking solicitor, of whom there were many in Mother Russia. The form was perfect, dictated most likely by an experienced hand. It contained worthy want, due respect, the moderated grief of an unfortunate widow, and it concluded with the most loyal feelings for the sovereign, and the signature of Praskovya Lobachevskaya written in two lines, thus exhibiting extra propriety and extreme respect. But who was this Praskovya, how and by what means did she make ends meet? No one knows.

So it was that on November 17, 1802, the three boys, Alexander aged 11, Nikolai 9 and Aleksei 7—all Lobachevskys—were matriculated at the gymnasium at government expense.

Careers were thus opened to the Lobachevsky boys.

There were not many in the old Russian Empire that achieved as much as Nikolai Lobachevsky did.

Of course there were much more brilliant careers. In beds of Russian Queens—Elizabeth, Anne, Catherine the Great—sturdy peasant or commoner lads (the genealogy was irrelevant in such cases, personal merits were more important) could easily earn the Privy Councillor or Chancellor rank and the estates with dozens of thousands serfs. Or under the crazy son of Catherine, the Emperor Paul, when one could skyrocket in a single fortunate instant from valet to count.

But for a man of science, the administrative career of Lobachevsky was extremely outstanding, though not unprecedented. And if we add that he proceeded by unsullied pathways, did not dissemble or seek, hardly cringed or flattered for promotion, he was indeed a rarely lucky minion of fortune.

True, he was no angel. He was a forward-thinking person of the times, nothing more. There were doings he was ashamed of—it was no easy task to serve and still to be morally immaculate in those days. He lived a complicated life and earned in full measure what is allotted to human beings in this world.

For the most part, an untroubled carefree youth, yet there were grievous losses too. The joy of success and the delight of creative work, yet dangerous unpleasantness in his student years. A radiant, cerulean—to begin with—scientific career, yet vicious, humiliating, taunting attacks of his enemies. Varied administrative and social activities, and the intrigues of his colleagues. Exalted praise for his administrative work, and pin-pricks to his self-esteem. Recognition by Gauss himself and the pleasant vanity of awards, yet the bitterness of offences. But again the love emanating from a happy family life.

At the end, destiny delivered old age, troubles at work, the death of his beloved son, the nervous breakdown of his wife, illness, blindness... and also, to the very last days of his life, an incomparable joy derived from his studies.

Chronologically, he spent 1802 to 1806 at the gymnasium with obligatory studies including:

Russian grammar and literature, history and geography;

arithmetic, algebra, geometry, trigonometry;

mechanics, physics, chemistry, hydraulics, surveying, civil architecture;

logic, practical philosophy;

and the foreign languages—German, French, Greek, the inevitable Latin and Tatar as well.

Then came military studies, which included tactics, artillery science and fortification.

That was not all. Add law, followed by such required items of society life as fencing, drawing, dancing and music.

Such was the course pursued in five years (not eight or ten!). Not everyone could cope with it. The Lobachevskys did. Apparently, they knew that for them there was no other way up. What is more, it was easier for them, for they were three brothers together.

Nikolai was the mischievous one. He was good at his studies. An ordinary capable child, the only difference from other, softer, easier-going boys of the nobility being his harshly practical approach to all things. It might have been the adult realization of the necessity to get ahead.

Among their teachers were cultured, talented people, some even out-standing. This especially concerns the mathematics teacher, Kartashevsky.

Then came January of 1807.

After some unpleasantness with Latin, Lobachevsky was accepted into the university. He was 14 years of age.

The first heavy blow came in July of 1807 when his beloved elder brother Alexander was drowned.

The result was a nervous breakdown for Nikolai, hospitalization, and a firm resolve to become a doctor. For over two years he studied medicine. True, he was the first in mathematics at the university, but his firm decision was that mathematics was not his vocation. The boy wavered between "duty" and vocation (remember he was only fifteen), deeply depressed by the death of his brother. He was obstinate, hard to get along with, though he was quite normal, very decent, and rigorously adhered to the student code of honour. He delved in all things that students do: fancy balls, the theatre, fights, and just pranks (for instance, once he rode up to the university building on the back of a cow; that is the episode that so many of his biographers claim to indicate a spontaneous protest against reaction).

Actually, the situation did change for the worse at the university during these years, and Lobachevsky's personal life was poisoned by a capable but rather unpleasant, unprincipled and ambitious Pyotr Kondyrev.

But Lobachevsky—it must be said—was not above ordinary boyish stupidity either, the same kind so often met with in hundreds of thousands of ordinary irritable, arrogant, quick-tempered boys so sure of themselves, positive that they can get out of any fix.

Kondyrev nearly ruined him. Lucky that the foreign professors Bartels, Littrow and Bronner who had been invited to teach at the university, were able to rescue the boy.

I say rescue because the issue was one of sending him off to the army as a soldier. In the best traditions of people of that sort, Kondyrev accused Lobachevsky of atheism and almost subversion of the establishment. It is not clear whether Nikolai was actually an atheist, but we do know that he never liked hypocrisy and the clergy.

Most likely, at that time and later Lobachevsky was "moderately progressive", with largely humanitarian views.

To extricate himself, he had to repent. Make a speech, express loyal sentiment, admit his mistakes and condemn them, and promise that in future he would not....

So much for pranks. But it was during these years that Lobachevsky finally made up his mind concerning his future: he would become a mathematician. He succeeded greatly in this field. He was the first mathematician of Kazan University, and Bartels was always glad to point out his attainments and talents.

If one recalls that in those years the whole of Russia had several thousand students, it isn't too much to say that Lobachevsky was already known throughout the country.

Kazan University was then an important centre of learning of the Russian Empire. If looking for contemporary analogies, one can compare it with the Siberian Division of the Academy of Sciences of the USSR. And Lobachevsky was the most promising young scientist there.

In August of the year 1811, at the age of eighteen Lobachevsky received his master's degree. His first success and the beginning of a number of very good years of intensive work. Socially, too, he was a success, accepted in the "best society" of Kazan. A young man about town, always well dressed.

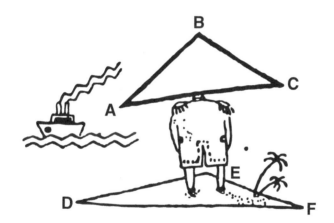

The war with Napoleon hardly touched him. His younger brother, Aleksei, tried to run away to the army, but was returned. True, Nikolai was worried sick until they found him. Lobachevsky, it must be said, always had very strong family feelings.

His moral code was already set—it was that of decency, decency in the meaning of the notions of that time.

Bartels, a very cultured teacher but a mediocre mathematician made him study the classics of science. Lobachevsky's only serious drawback, it would seem, was his excessive excitability and conceit. He was characterized once as "excessively self-centred". On the other hand, Lobachevsky clearly and soberly saw that he was yet far behind the great mathematicians of his days.

In March, 1814, he was made junior scientific assistant in the field of physico-mathematical sciences.

He began delivering his own lectures.

In July 1816 he was installed as extraordinary (associated) professor. This at the age of 24. His career had begun. Meanwhile the university was a beehive of intrigue with changes occurring constantly. For a short time, the reactionaries were on top. Then the "progressivists" got the upper hand. Trustees changed. In a word, then, life at the university was on "an even keel".

Lobachevsky had enemies in the reactionary party and he had influential patrons.

This was the period when Lobachevsky began to get interested in the problem of parallel lines. The beginning was standard. He attempted to find a proof. In 1815 at lectures he even described to his students the proofs he had found. But obviously he soon found his mistake. A rather elementary one too.

A big change came in 1819. This was the period of reaction throughout the country. It affected Kazan as well. The new trustee Magnitsky was a clever but unprincipled, cruel, cold climber-go-getter.

He was one of those people who fight their way to the top elbowing, climbing, pushing others out of the way and trampling on those who have fallen. His one and only aim was to reach the top. If reforms were needed, he would carry them through; if extreme obscurantism was the word, he would be the extremist. But he was a rather clever man with a flair for administration. He made his first appearance as an inspector summing up the situation in these words: close the Kazan University because of the free-thinking and general moral degradation.

Alexander I, however, decided not to destroy it, but to rectify the situation, and he put Magnitsky in charge.

Those were dark days for the university, but Magnitsky was kindly disposed towards Lobachevsky at first. He was possibly thinking of making him one of his protégés. During the years 1819 to 1821 Lobachevsky was on the upswing, elected dean, was responsible for the university library collection and member of the construction committee. Posts and titles came one after the other.

In February of 1822 he was elected an ordinary (full) professor. These were years when Lobachevsky acted against his conscience. True, with a person like Magnitsky there did not seem to be any other way out.

Bear in mind, too, that Lobachevsky was an independent-thinking person, quick-tempered and simply a hard person to get along with. Also his convictions were far removed from those of Magnitsky. I mean, removed under the situation of that time, because if suddenly liberal Lobachevsky's views were approved up above, by those in power, then in a moment... Magnitsky would turn out very progressive. All this boils down to the fact that in 1822–1823, Magnitsky was no longer kindly disposed.

In 1823 came the first major trouble in line of duty. His newly written textbook *Geometry* was rejected by Academician Fuss. It may be that Fuss was on the whole not right, although serious investigators agree that there were essential defects in the book and some of Fuss' remarks were quite true. Lobachevsky was stung to the quick and haughtily refused to reply to any of Fuss' remarks, or to correct any of the faults of the book, or even to take the manuscript back. His arrogance occasionally made matters worse for him. However, he continued to work intensely, and during these years he became fully convinced of the impossibility of proving the fifth postulate, and fully convinced of the equal rights of a non-Euclidean system of geometry.

A few pleasant events occurred in 1825 and 1826. Lobachevsky was put in charge of the construction committee of the university, he was also elected Chief Librarian of the University. His salary was raised to four thousand roubles a year. Very good money in those days.

Then Magnitsky was removed. It is curious that this was associated with the Decembrist uprising on December 14, 1825. Magnitsky had not been able to fathom the new situation after the death of Alexander I, he had risked everything on a big jump up, figuring that this was the time, but failed. He had placed everything on Konstantin, whereas Nikolai won out. And then it was that the old memorandum came to light where he had, it turned out, complained of the liberalism—no less—of Nikolai Pavlovich, then Grand Prince. Only in Russia could a paradox like that take place.

Quite naturally, an investigation was ordered. Certain sums of money, it appeared, had disappeared. It wasn't as if the barracklike set-up at the university was alien to the spirit of Tsar Nikolai; simply Magnitsky had overplayed his hand, and what is more important, he had simply not been able to guess the events of December 1825. He had lost. First discharged from all posts and then exiled to Revel[2] with an additional investigation

---

[2]Now Tallinn—A.S.

assigned into the money that had vanished.

Of course, the university, and Lobachevsky as well, rejoiced.

This is time to stop. February 23, 1826.

Up to this point we have witnessed the career of an interesting, gifted, pleasant provincial mathematician, though one not devoid of drawbacks. We have looked kindly on his climb. There has been no excitement, and we have not been unduly enthusiastic. A very decent career, where the hero was promoted from rung to rung of the ladder. He was not indifferent to his advancement and with the years his worldly wisdom grew, and escapades of his youth, ridiculous wild protesting of the malcontent, all remained in the past. Gradually, bit by bit, the common sense so usual in successful men accumulated. At the age of thirty-four he was a moderate man of fashion, a bit condescending in manner. Further advancements were in sight. Within a year he would be appointed rector (June 30, 1827)...

But on February 23, 1826, all these had become mere trifles of life, rather essential, but not over much so, and of course not decisively so.

That was the day the great mathematician gave his talk on non-Euclidean geometry to an indifferent, bored audience who understood nothing at all. Of course, if an angel had appeared and if there had been some sign from heaven—"This is the Man"—things might have been different. It might have been forgotten even that two days earlier an investigation of Magnitsky's activity at the university had begun. But in the absence of the angel, the very last thing that could have aroused his audience was a discussion by the very revered Nikolai Ivanovich on the theory of parallel lines.

Only Lobachevsky himself realized at that instant that this was the moment of triumph.

The lecture was forwarded to a commission for a review to decide on whether it should be published or not. The commission did not understand anything and, apparently, did not express any view at all. Either they did not want to endanger the well-being of a colleague, or there was some other reason. Anyway, the work was not published.

Then came 1827.

The new trustee was Musin-Pushkin. In many respects he was also a tyrant and ignoramus. But Lobachevsky had long been acquainted with him, and Musin-Pushkin judged that Lobachevsky was just the person to rehabilitate the university that had fallen so low under Magnitsky.

On his suggestion, Lobachevsky was elected rector, which post he occupied until 1846.

He was reelected six times, first by slight majorities and then by over-whelming majorities. That was something, if one recalls the atmosphere of constant intrigue within the university. There can be no question that he was a magnificent rector, who put a great deal of energy and love into his work, and a forward-thinking and very skillful administrator. He actually recreated the university. With great professional skill, he headed the construction work, set up a library, organized the regime of the students, and adjusted relationships between the Russian and German professors teaching at the university.

It is hard to see when he found time to devote to science. Yet all his basic scientific results were obtained during these very years of administration as rector.

The year 1829. The *Kazan Vestnik* (Kazan Herald) published his memoir *On the Principles of Geometry*. This was the first systematic description of non-Euclidean geometry.

The year 1830. In this year Lobachevsky became the hero of Kazan. Cholera hit the city. That was the terrible epidemic that swept across the whole of Russia and induced the poet Pushkin, who had to stay in quarantine in his estate in Boldino, to write *The Feast During the Plague*. Later, Pushkin admitted that he did not quite distinguish at that time between cholera and plague. The epidemic was extremely severe. In those days no one knew how to protect oneself against the contagion.

Lobachevsky arrogated unto himself the authority of dictator. The whole staff of the university, together with their families, were isolated from the rest of the world within the walls of the university buildings. Food was delivered with great care. Out of 560 persons, only 12 were taken ill (just over 2 percent) and they were immediately isolated. A brilliant result!

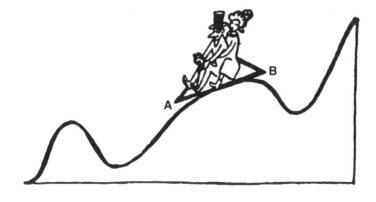

Then came the year 1832, when he married a nice young girl by the name of Varya Moiseeva. Love was mutual, though on his part a tiny bit too unruffled and somewhat too rational.

Outwardly, the years 1827–1834 were very lucky ones for Lobachevsky. Fortune was on his side. He was mature and everything was turning out the way he wanted it.

His activities during the epidemic were marked by higher titles and by the Tsar himself. Lobachevsky, though a civilian, displayed the efficiency and courage of a military leader; these were things Tsar Nikolai valued.

He simply must be rewarded. And His Imperial Highness rewarded him for his efforts—with the diamond ring, the title of State Councillor, the Order of St. Stanislav and the personal gratitude of the sovereign himself. Very good indeed. The track record of, now, State Councillor N. I. Lobachevsky was brilliant.

The first severe blow came in 1832 when Kazan University sent Lobachevsky's memoir *On the Principles of Geometry* to the Academy of Sciences for a review. The oral review was to be given by Academician Ostrogradsky. He took his time about it, and then stated: "What is true is not new, what is new is not true. The memoir is not worthy of the attention of the Academy of Sciences."

From that time onwards, Ostrogradsky became a sincerely vicious and implacable scientific adversary of Lobachevsky. Time and again he gave blistering reviews of Lobachevsky's work, because to him one thing was clear: that Lobachevsky was a provincial charlatan who must be driven out of science immediately.

Ostrogradsky was a good mathematician in the full meaning of the word, though his merits have been blown up unduly. He cannot, of course, be compared with such Russian mathematicians of the 19th century as Chebyshev, Markov, to say nothing of Lobachevsky. But if he really had wanted to, he could have made sense out of Lobachevsky's memoir. True, Lobachevsky himself was partly to blame. The style of his paper made reading it an arduous task. Not only is it concise beyond measure, but also not clear-cut in the least. True, Ostrogradsky should have been able to grasp the main idea. But he didn't, he was enraged and did not confine himself to an oral response.

In 1834, a well-known journal put out by Faddei Bulgarin, entitled *The Son of the Fatherland* carried an article in which both Lobachevsky as a scientist and his work were slashed to pieces. Today it appears to be firmly established that this "review" was inspired by Ostrogradsky.

But I believe that the article presents by itself a considerable interest and has instructive value. It deserves our close study.

The awful thing about it is that for the non-professional and even for the professional it is rather convincing. It would be hard to find a better example of the demoniacal power of demagogy, the force of conviction not via logic or reasoning but by implication, by intonation, sophistry and dishonest tricks of rhetoric.

The crude, simple, cheap humour that permeates the article is so convincing, acts so surely on the subconscious that it compels the reader to believe that this Lobachevsky is an ignorant self-satisfied nonentity. This is almost as much as spelled out in full by the author, who was without doubt a gifted writer. One finds it hard to find a more brilliant instance of the complete triumph of self-confident superficiality and idle twaddle over genius.

It starts at comparatively low gear.

"There are people who after reading a book complain that it is too simple, too ordinary, and contains nothing to think about. To such readers I recommend the Geometry of Mr. Lobachevsky. Here indeed is something to think about. Many of our first-rate mathematicians have read it, thought about it and still do not see the point. After that I hardly need say that I, having thought over this book for some time, did not get the point too. Frankly, I have understood nothing. It is even difficult to understand how Mr. Lobachevsky was able to concoct out of the simplest and clearest chapter of mathematics that we know geometry to be—how he could build such an abstruse, murky and impenetrable theory, if it were not that he himself helped us by saying that his Geometry differs from the *common* kind that we all studied and which, most likely, we cannot unlearn, and is only an *imaginary* geometry. Yes, that makes things clear indeed.

"Just try to picture what a lively, yet monstrous, imagination can conjure up! Why, for instance, not try to imagine black to be white, round to be quadrangular, the sum of all the angles in a triangle to be less than two right angles and one and the same definite integral to be equal first to $11/4$, then to ∞? Very very possible, yet to normal reason it is enigmatic".

How subtly journalistic. How neat is that "lively yet monstrous imagination". But that was only an introduction, kind of reconnaisance by fire, the heavy artillery was to come later. And the most powerful sort of guns are the rhetorical questions.

"But one asks why write such ridiculous phantasies, why have them published? I admit the query is hard to answer. The author did not once even hint at why he was publishing his composition, so we perforce must conjecture on our own. True, at one point he states clearly that, as he claims, the drawbacks which he had detected in the geometry so far in use compelled him to compose and publish this new geometry; but this, quite obviously, is untrue, and, in all likelihood, was said only to conceal better the true aim of his composition."

After that artillery barrage of sarcasm, we are ready for the direct assault.

"And to this allow me to add a few words about the man himself. How can one think that Mr. Lobachevsky, professor of mathematics in a university, would write seriously a book that could hardly be an honour to the lowest village school teacher? Every teacher should have common sense even if he does not have much learning. Yet this new Geometry is devoid precisely of common sense. Taking all of this together, I find it highly likely that the true aim of Mr. Lobachevsky in composing and printing his Geometry was a joke, or better, a satire on scholarly mathematicians, or perhaps on all scholarly writings of the present time. Thereby I do not merely tentatively suggest, I am fully convinced that the insane passion to write in a bizarre and obscure manner, which is so common of late among many of our writers, and the impudent desire of certain people to discover something new when their gifts are hardly enough to properly grasp what is old, are the two defects which our author wished to depict, and which he depicted with consummate skill."

Now that is what I call real writing; in complete keeping with the style and traditions of Faddei Bulgarin, himself a dashing daring "gangster of the pen". But one should not overdo it; it is time to display some kind of scientific approach. One should not allow the reader to have any doubts at the decisive turn in the battle. The operation begins with a somewhat risky admission. But the experienced warrior is apparently quite sure of himself.

"Secondly, the new Geometry, as I have already had occasion to state, is written so that the reader cannot understand anything. Wishing to get more closely acquainted with this composition, I concentrated all my attention, focussing every effort on every sentence, every word, every letter even, and for all that I dispelled so little the murk that envelopes this composition so completely that I am hardly in a state to relate to you what the matter is about, to say nothing at all about what is said..."

This is only an apparent retreat, for the question comes naturally, "If you understood nothing, then why do you undertake to reason and judge?"

No! He did understand what the matter was, but this was simply a manoeuvre to demonstrate to his co-readers how hopeless and monstrously disfigured was the construction of the enemy. And also to show his objective approach. Just look! "Would you like to see for yourself what the original is like?" Then followed a long quote from Lobachevsky's memoir. He knows how effectively precise is this manoeuvre.

Aside from the fact that the memoir was written in a ponderous complicated style, to comprehend Lobachevsky's ideas required a high level of mathematical culture and a concentrated and unprejudiced effort on the part of the reader. What is more, no isolated quotation permits one to judge the merits of a scientific work. More yet, after such a psychological build-up, an excerpt lifted out of context can be completely disarming. That was a sure move to capture victory. One last effort.

"But I must apologise, I simply cannot copy every word of it, for I have already said too much. And I cannot relate this matter in brief, for that is where the most incomprehensible begins. It would seem that after a few definitions, composed with the same art and the same precision as the preceding ones, the author says something about triangles, about the dependence of the angles in them upon the sides—therein lies the difference between his geometry and ours—he then proposes a new theory of parallels, which—and he admits as much—nobody is capable of proving whether it exists in nature or not; finally, this is followed by a consideration of how, in this imaginary geometry, one determines the magnitude of curved lines, of areas, of curved surfaces and volumes of solids. And all of this, I must repeat once again, is written so that nothing at all can be understood..."

The amusing thing is that though the author mastered the titles of Lobachevsky's theorems, he was not up to grasping the fact that Lobachevsky's geometry differs from ours *solely in the theory of parallel lines*. But why indeed should one need to understand anything? The enemy is in retreat, completely routed, all that is needed is to consolidate the victory.

"... Mr. Lobachevsky deserves to be praised for taking upon himself the labour of explaining, on the one hand, the arrogance and shamelessness of pseudo-inventors, and, on the other hand, the naive ignorance of the admirers of their pseudo-inventions. However, realizing the full value of Mr. Lobachevsky's composition, I cannot refrain from blaming him slightly for not giving his book a proper title and compelling us to cogitate so wastefully for such a long time. Why, instead of the title *On the Principles of Geometry*, could he not have named it, say, *A Satire on Geometry*,

A Caricature on Geometry, or something of that nature? Then anyone would immediately see what the book was about and the author would have avoided a host of unpleasant interpretations and arguments. It is lucky that I have been able to penetrate to the true purpose of this book, or heaven knows what people would think about it and its author. Now I think and am even convinced that the worthy author will feel greatly obliged to me for having demonstrated the true point of view that one must take when reading his composition...."

This lampoon is quoted more or less in full in every biography of Lobachevsky. However, though the biographers are indignant and abuse the writer in every imaginable way, they usually lose sight of the most important thing—the fact that it is a very cogent piece of writing. I am not in the least interested in the one (or several) who wrote it; theoretically one can even assume that he was sincerely fighting for the purity of science.

But, too, one can readily see what the reaction was of the readers and also what this article costed Lobachevsky.

After reviews of that kind, people take to their beds, give up work altogether, or even commit suicide.

On the background of this pamphlet, Gauss' letter to Bolyai is that of a tender, loving solicitous father. Taurinus—another one of Gauss' "victims"—burned his paper for the sole reason that Gauss, offended, dropped the correspondence.

Outwardly, this story seemed not to have involved Lobachevsky at all. He reacted with an amazing lack of spirit. There were a few questions, and a year later he published, in the transactions of the university, a very calm and restrained reply. Too, an extremely cool answer was sent to the *Son of the Fatherland.* Faddei, of course, never published it. And Lobachevsky seemed not to care. He never tried to insist. That was the end of that.

It would be wrong to think that Lobachevsky was not a man of action. His whole life and his 19-year-long tenure as rector demonstrated quite the contrary. Apparently, in this particular instance he considered it below his dignity to enter into a discussion. In general, he was surprisingly indifferent to any popularization of his ideas. This is a psychological riddle because in all other things he was an extremely practical man. What is more, he had it in his power to put an end to the outpourings of his adversaries.

In 1840 he published one of his works in German. And already in 1842 he was elected—on the suggestion of no other than Gauss himself—to Corresponding Membership in the Göttingen Royal Society.

Gauss read Lobachevsky's paper and was carried away by it. True, carried away in his own particular way. There followed opinions full of admiration expressed in letters to his friends; then very sharp replies with respect to a review of Lobachevsky's work given in a German journal. Essentially, this review was of the same nature as the pamphlet published in the *Son of the Fatherland*, and Gauss' description of the reviewer was very harsh. Finally, in letters to his Russian correspondents he constantly inquired about Lobachevsky and even asked to convey his greets to the Russian mathematician.

But there was not a word in the press, not a single letter to Lobachevsky himself, with the exception of the strictly official correspondence pertaining to his election. True, he had wanted to write and ask for reprints of Lobachevsky's works. That is, he was on the verge of doing it, but he never wrote.

Well, all right, Gauss had his own reasons. But how are we to account for Lobachevsky's silence?

After being elected Corresponding Member, he was of course quite positive that Gauss had read his paper and approved of it. There can be no question that such recognition was extremely important to him and very heartening. What would be more natural than to send Gauss his papers or at least to write him a letter asking for an appraisal of his ideas?

And this is to say nothing of the fact that if Lobachevsky had ever received such a response from Gauss, then the professors of Kazan University and the whole Academy of Sciences would straightway repudiate all earlier attacks and would rejoice in recognizing Lobachevsky as the greatest mathematician of Russia.

Suppose he was totally indifferent to the opinions of those around him, though that is very hard to imagine. Even so, he himself should surely have been interested in a detailed appraisal of his work by Gauss.

He never wrote such a letter to Gauss. Why? Modesty? Pride? The fear of appearing to be importunate? I do not know.

It may be that he was deeply offended because of Gauss' attitude, for surely the great man could have written a couple of encouraging words to a Corresponding Member of the Göttingen Society. It may be...

I simply cannot find a satisfactory version, even ever so slightly. The only thing to be said is that this mysterious chain of events demonstrates what a complex and uncommon man was Lobachevsky. For, beginning with November of 1842, he undoubtedly realized that recognition as a mathematician in Russia could come to him at any moment that he himself desired. But he did not write. When a matter of his life is at stake, he is so chastely restrained! Lobachevsky the mathematician was quite a different man from Lobachevsky the university rector. The mathematician was impractical, reserved, philosophically placid.

All these years he worked intensively striving to find a rigorous proof of noncontradictoriness.

Quite separately from this flowed his duties at work, his family life, his ups and downs in day-to-day life. His wife proved to be of a serious turn of mind, and quibbling and open scandals occurred fairly often in their home. And through it all he was the model stoic. "Oh, my dear Varvara Alekseevna..." with all respect—and then he would disappear into his fortress, his study. Or he would sink into silence puffing at his pipe.

There were a lot of children in the family. He seemed rather indifferent to the girls but he loved the boys with a kind of jealous, harsh, carping love. Particularly Aleksei, the eldest. So capable, so much like himself in childhood.

Meanwhile there was no end of administrative duties, which he performed in model fashion, running the university efficiently. And do not forget the difficulties of the times. The government and the Tsar were satisfied.

On several occasions, His Imperial Majesty expressed his grace and benevolence to His Excellency the Active State Councillor. And in the offing lingered the still higher rank of Privy Councillor.

Money matters were not in the best possible order, but he was still full of energy and not too old.

Of course, there were endless intrigues and smearing and muck-raking among his colleagues. But so always has been the case. He took them in his stride, became severe, reticent, pedantically composed. But such traits are common to aging men. He was ordinary in all things and habits. His Excellency was a good host and knew a thing or two about cooking.

At the club there was card-playing—he liked preferans.

But his recreation more often consisted in translating from the Greek and Latin.

He loved his university, and the students loved him. His work occupied him completely.

Everything was typically Russian. His brother Aleksei was a heavy drinker. A relative of his wife was a gambler who lost a large sum of Lobachevsky's money. His sons were grown up now, students. His favourite one gladdened his heart, the younger one didn't; he was so obviously no mathematician.

What was on his mind all these years? What gave him the strength to pursue the study of his geometry so persistently? How was it possible to carry on with his geometry through all the vicissitudes of a lifetime and not to turn into the most ordinary of state councillors? Where did he find the will? What buoyed him up all these years? What were his thoughts when he was alone in his study? What were his dreams? And hopes?

I have no answer, and no one will ever have, I'm afraid. Nikolai Ivanovich Lobachevsky appears to me as one of the most mysterious men in the whole history of science.

In the opinion of many of the most cultured people of that period, Lobachevsky was, on the whole, a very respected official. He was also "an eccentric practically out of his mind", "the mad man from Kazan".

Of course, his real life began behind the doors of his study. Quite naturally. But what maintained him, what concentration of will-power, what driving force? What was he guided by—love, hatred, hope, superciliousness, habit turned to instinct? I cannot say. I'm afraid no one can say. Because all the riches of the archives add nothing about this second, most important life of his, the life that began when he was alone with his computations. Perhaps there is, after all, just one thing that opens him up a crack.

In the year 1853, his most dearly loved boy Aleksei died. Within a few months Nikolai Lobachevsky was a sick man, broken. He began to lose his sight, and the illness progressed rapidly and implacably.

He had three more years to live. His routine went on and he still performed his duties, but life was already gone.

And when he recalled his efforts to make his son study mathematics; how, though self-controlled and calm most of the time, he would shout abuse when the boy was lazy, or would rejoice majestically when he came to his room to find the boy celebrating a successfully passed exam with his friends: "Continue, gentlemen, I shall not bother you"; recalling all these things, one may conclude that this harsh, unsociable man was probably kept alive by a single romantic dream—that of his son continuing his Geometry.

The death of Aleksei meant that he himself was dead. Misfortune does not come unaccompanied. During the last three years of Lobachevsky's life, one calamity followed another.

Perhaps he was now to some extent immune, for the end had already come. There remained only one thing, his Geometry.

Already blind, with only a few days of life left, he dictated the last of his works.

Chapter 9

Non-Euclidean Geometry.
Some Illustrations

Let us look into the curiosity shop of non-Euclidean geometry.

Well, they seem to be curiosities only because our intuition, firmly rooted as it is in Euclidean notions, will resist at the first steps the notions of the new geometry.

To compel our sensations to vote for theoretical equivalence of the two geometries, one has to put intense and sustained efforts into the study of the geometry of Lobachevsky.

Only then, what at first superficial glance appeared absurd and paradoxical, will begin to shine with the calm cold beauty of logic and truth.

Since we speak of beauty, let us recall an analogy from the arts.

The canvases of the impressionist painters that give us such great enjoyment today were derided with guffaws and shouts of disgust when first displayed in the art salons at the end of 19th century. This reaction was of the same nature as that of Lobachevsky's contemporaries to his works. One can make at this point a simple general remark: unfortunately, it is still a revelation to most people that one should better try to understand the issue at hand before giving an opinion.

Too often, fragments of distorted, twisted information that reach us by accident are taken as sufficient grounds for authoritative assertions, no matter whether they are for good or for bad. Incidentally, the geometry of Lobachevsky was once again luckless in this sense in a most amusing fashion.

Quite some number of years ago, I came across the following phrase in the works of a very well-known writer: "Lobachevsky proved that lines that are parallel according to Euclid intersect at infinity." This was then followed by a round of clever, sweeping generalizations. I don't remember exactly what about. Almost about what I am now writing.

By the same token of that penchant for superficial reasoning that I have just noted, I decided that the author had never really heard of Lobachevsky's geometry. But this same phrase cropped up so constantly in articles and books by other writers that I realized, one fair day, that the phrase acquires sense if the "parallels" are understood in the meaning of Lobachevsky! Just a couple of pages from now we shall see that these lines are by no means Euclidean parallels. They are related, roughly, like sea pilots of the Middle Ages and air pilots of today. The single term used to denote two different notions created confusion in the minds of people foreign to mathematics. Perhaps they do not deserve to be harshly censured, yet neither do they warrant any encouragement.

Actually, I seem to have found the primary source of the "literary version of Lobachevsky's geometry".

The culprit seems to be Fyodor Dostoevsky! In his *Brothers Karamazov*, Ivan explains his moral and philosophical credo to Alyosha and, among other things, has this to say: "But you must note this: if God exists and if He really did create the world, then, as we all know, He created it according to the geometry of Euclid and the human mind with the conception of only three dimensions in space. Yet there have been and still are geometricians and philosophers, and even some of the most distinguished, who doubt whether the whole universe, or to speak more widely the whole of being, was only created in Euclid's geometry; they even dare to dream that two parallel lines, which according to Euclid can never meet on earth, may meet somewhere in infinity. I have come to the conclusion that, since I can't understand even that, I can't expect to understand about God. I acknowledge humbly that I have no faculty for settling such questions, I have a Euclidean earthly mind, and how could I solve problems that are not of this world?"

I do not think of identifying Dostoyevsky with Ivan Karamazov, and we are not here dealing with the problem of the existence of God. But when it comes to geometry, this is the reasoning of Dostoyevsky himself. And the fact that it is a magnificent piece of writing just shows how a shallow, superficial intuition on the part of a superficial dilettante is so unwittingly elevated to an absolute principle. Strictly speaking, there is not a single correct idea in the whole passage. This is all the more exciting since the magnificent and purely analytical mind of the author is also apparent in every word.

Ivan then brings his intellectual eccentricity relative to geometry to its logical culmination and even extends it to the realm of physics:

"Even if parallel lines do meet and I see it myself, I shall see it and say that they've met, but still I won't accept it!"

Quite naturally, I do not intend to draw any far-reaching conclusions (any at all!) about Dostoyevsky's writing from these excerpts. Ivan Karamazov had no interests in geometry of course. For him the whole thing was simply an illustration of his ideas.

But for us this can serve as an illustration of a distorted conception of science and of light-minded reasoning about uncomprehended things (and of course the obvious obscurantism of Ivan Karamazov).

One might, however, justify Dostoyevsky at least in the sense that he did not perhaps make any actual mistake. The names of the geometers are not given, and so one might hope that Ivan was describing elements of Riemannian or projective geometry.

However, since on the one hand, the words "non-Euclidean geometry" are associated with the name of Lobachevsky, and, on the other hand, all cultured writers have undoubtedly read Dostoyevsky with care, the words of Ivan were consistently extrapolated to Lobachevsky himself.

All these literary-psychological explorations are useful, aside from general ideas of a didactic nature, in that they help us to grasp the intellectual courage of Bolyai and Lobachevsky.

Having made this digression, let us return to our curiosity shop.

We will naturally confine ourselves to only a few theorems and will not at all talk about solid geometry. Throughout we will agree that everything occurs in a single plane.

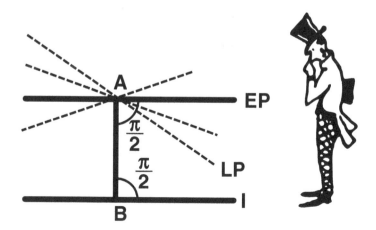

First, of course, the postulate of Bolyai-Lobachevsky or "Postulate Λ" the antagonist of Euclid's fifth.

Through a given point A not belonging to a given straight line I, it is possible to draw, in addition to Euclid's parallel, at least one more straight line that does not cross I.

Whence it follows immediately that one can draw an infinity of such straight lines.

Look at the figure at the preceding page. A perpendicular is dropped from point A to the straight line I. Euclid's parallel—line EP—is naturally perpendicular to this perpendicular.

The dashed line (LP) does not intersect I.

By means of symmetry reasoning (bend the drawing along the perpendicular AB!) it is clear that there will be another straight line of exactly the same kind. It is also dashed. Further, it is clear that any of an infinitude of straight lines drawn through A inside the angle between the straight lines EP and LP will not intersect the straight line I either. We thus have: *Through a given point not lying on a given straight line I, it is possible to draw an infinity of straight lines that do not meet I.*

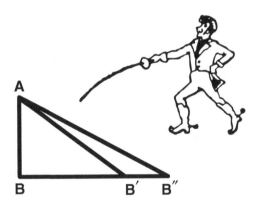

But one can naturally also draw an infinity of straight lines which do meet the given line. They may be drawn to any point (of the straight line) arbitrarily distant from the base. Let us take any point B' and join it by a straight line to A. This can always be done on the basis of a familiar axiom.

And so we have a straight line passing through both A and B'.

However, due to the continuity of the bundle of straight lines, there must be a boundary line that separates the two classes. This is either the last "intersecting" straight line that meets the straight line BB'', or the first "nonmeeting" line. It is readily seen that there can be no last "intersecting" line. Indeed, suppose it exists. Let it be AB' in our figure. But then, if we take B'' beyond B' and connect it with the point A, we get a new straight line lying beyond B' and intersecting the straight line I.

Consequently, the boundary straight line is the first one that does not meet the straight line I.

There are naturally two such straight lines: one for each direction. Within the angle formed by these straight lines, we can draw an infinity of straight lines that do not meet the line I; these will also include Euclid's parallel.

Lobachevsky gave the name parallel to these two extreme nonintersecting straight lines.

As you see, they do not have any relation to the parallel as understood by Euclid.

Stretching the point a bit (in Karamazov's direction), we may say that they, as it were, intersect the given straight line $B'B''$ at infinitely distant points. However, it is not at all clear what is meant by "infinitely distant points", so it is better not to use that phrase at all.

In Lobachevsky's terms, all straight lines within the angle "diverge" from the straight line I.

To summarize, then, relative to a given straight line there are three types of straight lines that may be drawn through any external point.

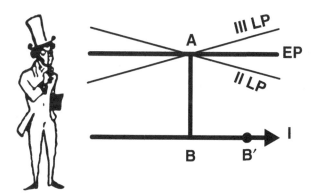

1. *Converging* (intersecting) lines; there is an infinity of them.

2. *Parallel* lines. There are two. Of each we say: parallel II is parallel to the straight line I in the direction BB'; parallel III is parallel to I in

the direction $B'B$. The meaning of these words is clear from the figure.

3. *Diverging* straight lines. These comprise the infinitude of lines within the bundle, one of which is the "Euclid's parallel".

Those are the terms. Now let us look into the theorems.

With regard to parallels, Lobachevsky demonstrated that they approach without bound a given straight line (without ever meeting it) and recede, without bound, on the other side.

So far, this is not such a strange result.

But the next one is like a thunderbolt.

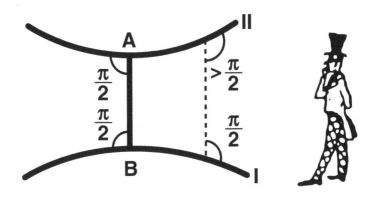

Two diverging straight lines always have a common perpendicular which is the shortest distance between them. They recede without bound on both sides of the perpendicular. This naturally holds true for the special case of "Euclid's parallels" as well.

Thus, a perpendicular dropped from any point of the straight line II onto the straight line I is, firstly, greater than the mutual perpendicular AB and, secondly, does not form a right angle with the straight line II.

This is indeed strange. But the proof is immaculate.

Accordingly, the locus of points equidistant from the straight line turns out to be a curved line.

These are only the first steps.

At this point, Lobachevsky introduced a new and very important concept: the parallel angle.

This is the acute angle between the straight line parallel to I and drawn through point A, and the perpendicular AB dropped from this point onto the straight line I. Thus, the parallel angle is CAB. According to Euclid, it is naturally always equal to $\pi/2$.

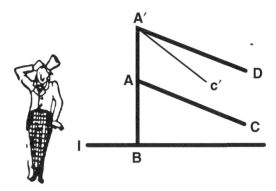

It will readily be seen that this angle depends on the distance between A and the straight line I, and diminishes with increasing distance.

Indeed, take a point A' on the prolonged perpendicular AB and draw from this point a "Euclidean parallel" to the straight line AC. It will intersect the perpendicular AB at the same angle as the straight line AC.

$$\angle DA'B \;=\; \angle CAB$$

But we know that from A' it is also possible to draw a straight line $A'C'$ parallel to AC in the sense of Lobachevsky.

The angle $C'A'B$ is also obviously less than the angle $DA'B$.

It is obvious that if the straight line $A'C'$ does not intersect AC, it will definitely not intersect the straight line I. It will either diverge from it or be parallel to it. (From here onwards I will no longer say "in the sense of Lobachevsky". We will adhere to his geometry and to his definitions.)

Lobachevsky actually proved the theorem:

If two straight lines are parallel to a third in one direction, they are parallel to each other in the same direction.

And so the angle $C'A'B$ is the parallel angle to the straight line I at the point A'.

The parallel angle is a function of the distance to the straight line. Lobachevsky denoted this function as $\Pi(x)$; x is the distance, that is, the length of the interval AB.

We have already seen that this function diminishes with increasing x. Lobachevsky also investigated its behaviour with decreasing distance x and showed that the parallel angle $\Pi(x)$ then tends to a right angle.

Symbolically, scientifically, this looks like

$$\lim_{x \to 0} \Pi(x) = \frac{\pi}{2}.$$

But if we recall that a right parallel angle corresponds to Euclidean geometry, then it is clear that at small distances the geometry of Lobachevsky is practically indistinguishable from the geometry of Euclid.

Clear so far. What is not clear, however, is what we mean by "small distances".

The words "small" or "large" have meaning only if we know what is being compared to. Without that knowledge they are devoid of any content. There should obviously be some kind of length, of standard that can be used for purposes of comparison.

How does such a standard enter here? It is well worth recalling Legendre at this point. He too discovered that the parallel angle depends upon the distance. Actually, all that needs be done (as we have already mentioned) is to analyse his proof with regard to the sum of the angles of a triangle. The very fact that a relationship like this appears seemed to Legendre so absurd that at one time he declared it the desired absurdum that proved the fifth postulate. Legendre's reasoning was ingenious. He argued more like a physicist than a mathematician.

Actually, he employed a very strong method of qualitative analysis of physical problems called the dimensional method. Brought up to date, his reasoning might look like this.

We see that the parallel angle is a function of only one paramer, the distance from the straight line. No other linear dimensions enter into the problem. We write $\phi = \Pi(x)$.

Now let us see what we have written. The angle ϕ is a dimensionless quantity. (In radian measure, an angle is the ratio of the arc of a circle to the radius.)

On the left we have a dimensionless quantity. It remains the same, no matter what units of measurement are used, whether centimetres, metres, inches or what have you.

On the right, however, the function is that of a dimensional argument. It makes no difference what form it has. The important thing is that no matter what it is, its numerical values will vary with the unit of measurement. If say, $\Pi(x) = 1/x^2$, then for $x = 1\,m$, $\Pi(x) = 1\,m^{-2}$.

But if the unit is 1 cm, then

$$\Pi(x) = \frac{1}{(100\,cm)^2} = 10^{-4}\,cm^{-2}.$$

This is obviously nonsense. The relation we have suggested is impossible. Consequently, the fifth postulate is proved.

The chain of reasoning is absolutely correct, but the conclusion is not. The conclusion has to be different. From the same arguments of dimensionality it is clear that in our formula there should be a nondimensional quantity in the argument of the function on the right hand side. This is what the equation should look like:

$$\phi = \Pi\left(\frac{x}{k}\right),$$

where k is some scale which we still do not know. The question is where do we find the scale k? In fact, the whole preceding analysis showed that the parallel angle ϕ depends on only *one* distance, the distance of the point from the straight line.

There is only one way out. We have to assume that in the new geometry there is a specific, nature-given constant unit. A kind of constant length that determines all other lengths.

This is strange but not entirely absurd. For instance, the two-dimensional Euclidean geometry of a sphere has such a fundamental length. It is the radius of a spherical surface. And so when employing the formulas of ordinary Euclidean spherical geometry for a geodetic mapping of Mars we will have to bear in mind that some of the "constants" of our terrestrial tables will undergo appreciable change.

Lobachevsky was not embarrassed by the apparent paradox. He introduced such constant scale k and found the equation for the parallel angle. It is so simple that we give it here:

$$\tan\frac{\phi}{2} = e^{-x/k},$$

where e is the base of natural logarithms.

From this equation we immediately see that when $x/k \to 0$, then $\tan(\phi/2)$ tends to $e^0 = 1$ or $\phi/2$ tends to $\pi/4$ and ϕ to $\pi/2$. When $\phi = \pi/2$, two Lobachevsky's parallels coincide between themselves and with the Euclidean parallel.

But x/k is close to zero when $x \ll k$.

Now what we said just a moment ago about small intervals has acquired precise meaning.

If the distance from the point through which we draw a parallel to a given straight line is much less than the constant scale k, then the geometry of Euclid is fulfilled in approximate fashion.

In the limiting case when $k = \infty$, Euclid's geometry is always fulfilled and with absolute precision.

The first question that naturally confronted Lobachevsky was how to find the scale k?

And here it turned out that his geometry was in a certain sense "better" than Euclid's. No theoretical arguments help to define k. It is what physicists call a "constant of the theory". It can only be found experimentally, by means of concrete physical measurements.

It is of course impossible to measure the parallel angle directly, but it is, for instance, possible to measure the sum of angles of a triangle. The "defect of the sum" in a given triangle depends on the value of k.

You remember that both Lobachevsky and Gauss urged such measurements but nothing came of them.

Generally speaking, Lobachevsky never said that it was precisely his geometry that describes the world. Quite the contrary, he inclined towards the view that in this world, it is Euclid's geometry that is accomplished.

But that is not so important. The remarkable thing is that from the very first steps the new geometry was closely tied in with physics and that it was inconceivable to disassociate it from experiment.

This naturally put forth the salient problem of the relationship of geometry in general to the real world, the possibility of different geometries in the real world.

It is not that such possibility was not sporadically discussed before, but for two and a half thousand years mathematicians took a dim view of it, regarding the whole matter as futile and absurd.

Willy-nilly non-Euclidean geometry revived this question. Are we indeed so sure that God made the earth in accord with the laws of Euclidean geometry, as Ivan Karamazov would have us believe?

A beauty of mathematics is that abstract mathematical formulas may engender totally unexpected ideas, which even their discoverer never suspected.

All these conclusions are so charmingly elegant that one can understand Bolyai and Lobachevsky who had faith in the logical rigour of their system.

Note also that we have discussed here only one of the conclusions of Lobachevsky's very first work, his paper of 1826.

Already at that time, he developed this scheme in depth and obtained also other results, which were no less beautiful. But in mathematics, beauty and faith is an important, but not decisive factor.

There were no guarantees that a logical contradiction might not pop up in the future.

Lobachevsky spent the rest of his life in persistent attempts to find this proof. He strived to demonstrate with complete rigour that his system was flawless. On the way he worked out a great diversity of the most unexpected consequences of his geometry, penetrating ever deeper.

In this respect, he is without a doubt head and shoulders above his contemporaries, for neither Bolyai nor Gauss covered the ground as well as he did.

He did not find the proof, though he was rather close to the basic idea.

But from human point of view, his persistent, never swerving struggle towards a single goal is worthy of our admiration.

Chapter 10

New Ideas. Riemann. Noncontradictoriness

No, this will not be a chapter about things of startling beauty. I will be honest with the reader. At least the first half will be rather dry mathematics.

First about the theory of surfaces. The progenitor was again the same old Gauss.

Let us imagine that on some kind of whimsically bent surface there reside intelligent *two-dimensional* beings. What will their geometry be like? Secondly, how will they be able to see that their surface is curved?

At first glance, the second question may appear quite naive. The reader may be recalling proof of the sphericity of the earth given in elementary-school geography books. Don't hurry, remember that we are *three-dimensional* beings living on a two-dimensional surface. The elementary-school proofs rely on this fact.

To rid ourselves of the illusion that this is simple, think over the question: How can one find out that our three-dimensional world is curved, and what in general does this so frequently employed phrase mean after all?

The three- and four-dimensional worlds will be looked into later on, for the present let us return to surfaces.

Gauss began by introducing a marvellous quantity that defines the geometry of a surface. It is called Gaussian curvature. The fundamental property of Gaussian curvature is: it remains constant under any bending of the surface so long as no stretching occurs. It is intuitively clear what this means, but a rigorous formulation is better: if in the bending of a surface there is no stretching, then, first of all, the lengths of all curves drawn on the surface remain unchanged; secondly, the angles between them remain the same too.

This can he stated somewhat differently. Take a sheet of paper. Bend

it. Then measure the Gaussian curvature at some point. Now you can do whatever you want to this sheet (except stretching or tearing it), like twisting it into the most bizarre forms, and the value of the Gaussian curvature at that point will not change. The Gaussian curvature is so important a concept that we will define it in more rigorous fashion. To do this, we will have first to find out what are radii of curvature at a given point of a surface.

We consider some point of a surface and draw a line normal to it. What is a normal? To explain we will need one more concept, that of a tangent plane. We give an almost rigorous definition. We consider all possible curved lines located on the surface and passing through a point P.

It turns out that the tangents to all these curves lie in one plane. This is not evident at first glance, but it can be proved rigorously. It is the entire collection of tangent lines that forms a tangent plane.

For the "cap" drawn in the upper figure on page 168, the location of the tangent plane is rather obvious. But sometimes the tangent plane is located more intricately relative to the surface (see the figure above).

Now let us define precisely the notion of a normal. The normal is a straight line perpendicular to the tangent plane. Choose a direction (*any* of two directions is OK) and define the normal vector as the unit vector pointing in that direction. We can now define the concept of principal radii of curvature. Pass a plane through the normal. There are clearly an infinitude of such planes. We take any one to begin with. A plane curve is formed by the intersection of the plane and the surface. One can always choose a circle that is contiguous to this curve near the point P. I shall not

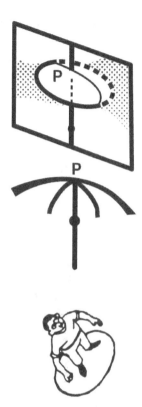

explain the exact meaning of these words in the hope that your intuition will suffice to create the proper image.

The radius of this contiguous (or *osculating*, as mathematicians prefer to call it) circle R is called the radius of curvature of the plane curve. It is positive if the chosen normal vector point in the direction of the centre of the osculating circle (and one can prove that the centres of the circles are always located on the normal) and negative otherwise. Since an infinity of planes can be passed through the normal, we get infinitely many radii of curvature, among which there is the greatest and the smallest one. It can be proved that plane curves to which the smallest and greatest radii correspond are mutually perpendicular at the point P. These two radii, R_1 and R_2, are called the *principal radii of curvature* of our surface at the point P.

If R_1 and R_2 are of the same sign, so that the centres of the principal osculating circles lie on one side of the surface, the point P is called *elliptical*.

If they lie on the different sides, we are dealing with a *hyperbolic* or a *saddle* point. Finally, there are *parabolical* points. They are points, where one of the principal radii of curvature is equal to plus or minus infinity.

The Gaussian curvature at any point of a surface is defined as:

$$K = \frac{1}{R_1 R_2}.$$

Now we can set our findings out in a table:

In the elliptical point	$K > O$
In the hyperbolical point	$K < O$
In the parabolical point	$K = 0$

Now let us see what properties the surface as a whole can have. Imagine some surface and try to cover it with a piece of closely adhering cloth. The rules of the game are: that the cloth cannot be cut or stretched, and has to cover the surface without any folds.

If a lady confronted a tailor with such demands, she would be dismissed without further ado, and he would be right in doing so.

The reader would do well at this point to stop reading and try to picture the properties that the figure of our hypothetical lady of fashion should possess. After what we have found out about the properties of Gaussian curvature, the answer is simple. The piece was flat at first. Which means the curvature was zero at every point. Bending without stretching does not change the curvature. This means that a flat piece of cloth may be bent only onto a surface whose curvature at every point is strictly equal to zero.

A cylinder, for example. It is easy to see that the Gaussian curvature is strictly zero on the lateral surface of the cylinder. Or, in other words, every point of the surface is parabolical. If you have mastered the concept of curvature, you will readily see that the second example of a suitable surface is the cone.

On the other hand, we cannot bend the plane onto a sphere as required. The curvature of a sphere is constant and positive. It is precisely this circumstance that causes cartographers so much trouble.

We must observe, rather tardily, that all along we have had in view only "good" surfaces. To put it crudely, "good" surfaces are those that have no sharp points or edges. The vertex of a cone, for example, is a "bad" point.

Also, when we speak of bending one surface onto another, we have in view, strictly speaking, the bending of a sufficiently large piece, but not the whole surface. To take an example, the entire lateral surface of a cone can be unfolded onto a plane only if we make a cut along the generatrix.

The last term we have to explain is a geodesic line. A geodesic is a curved line drawn on a surface between two points so that any other curve is longer. This definition is one of those "almost rigorous" ones, but I have hopes that only non-mathematicians will read this chapter and so there will be no one to criticize me.

Hypothetical beings of two dimensions who live on such a surface will say that the geodesic line is the shortest distance between two points. Incidentally, three-dimensional beings (like we are) would say the same thing if we impose the condition that they should not leave the surface.

To us earth dwellers living on a sphere, the shortest distance between two points on the earth is an arc of a great circle. It is precisely along the arc of a great circle that navigators sail their ships in making the briefest voyages.

There is also a different rather unexpected aspect of the geodesics. Now we said that a plane may be bent onto a surface whose curvature is constant and equal to zero. Or—what is the same thing—that such a surface may be unfolded onto a plane. Then any figure drawn on the plane will turn into a similar figure on our surface. But the angles between lines do not change during the bending process. Neither distances do. And one can show that the shortest lines on the plane—straight lines—will pass to the shortest lines on the surface—the geodesic lines!

Therefore, for a cylindrical triangle, for instance (naturally, its sides are formed by curved lines), the sum of the angles remains the same as in the plane triangle. We can go on reasoning in the same vein. To every geometric concept on the plane we can correlate a corresponding image on the surface.

It is rather easy to see that all the theorems that hold for the plane can be carried over without change to the surface. The only thing that we must bear in mind is that these theorems now hold true for the "images". If Euclidean geometry is accomplished on the plane, then it will be accomplished on a cylinder for the "images" as well.

We have now touched on one of the most remarkable and beautiful aspects of mathematics. So long as we are not interested in any practical applications, it is all the same to us what our theorems speak about. We only want them to satisfy the demands of logic. What is more, we do not even know what we are talking about. It is only the physicist that has to know what is "actually" taking place, what his world is really like.

For the physicist, a straight line is a ray of light. For the mathematician, it is one of the basic undefined concepts. There is no way of distinguishing between the straight lines on a Euclidean plane and the geodesic lines on the surface of a cylinder if they are compared solely from the point of view of axiomatics.

Let us conjure up a fantastic picture. Two two-dimensional worlds. One on the plane, the other on the surface of a cylinder. Intelligent beings live in both worlds. Suppose they have set up some kind of communication. The two-dimensional "plane" mathematician and the two-dimensional "cylindrical" mathematician would assert with great satisfaction that their geometries are the same.[1]

If the system of axioms were contradictory on the Euclidean plane, we would know immediately that it was contradictory on the cylinder as well.

One could explain to the other the theorems he has developed, and the latter could accept them without making any modifications. They could work together without the slightest friction. Now the physicists in the two worlds would probably be in conflict from the very start. They would claim, each in his own world, that the laws of nature are different in the other world.

Well, in case if a ray of light in the cylindrical world followed a geodesic line, they also would not be able to detect a difference.

The reader has by this time guessed that we are rather close to the problem of noncontradictoriness in non-Euclidean geometry. If we were able to find, in ordinary Euclidean space, surfaces on which Lobachevsky's geometry is accomplished... if these surfaces could be made so that the whole of the Lobachevskian plane could be mapped onto them, then the problem would be solved.

The first "if" is satisfied. Such surfaces (called pseudospheres) exist. These are surfaces with a constant negative curvature. But the second condition has us stumped. The entire surface of a pseudosphere corresponds to only a piece of a Lobachevskian plane.

Let us forget noncontradictoriness for a moment (we will come back to it at the end of the chapter!) and say a few words about Riemann. In the year 1854 this morbidly shy youth opened up fresh vistas in mathematics.

Go back to the Gaussian curvature. Translating our previous discussion about the lady of fashion and her cloth into pure mathematical language, we can say that the curvature is an *internal* property of the surface and does

---

[1] This is true as long as the sizes of the figures do not exceed the size of the cylinder.— A.S.

not depend on how this surface is *embedded* in three-dimensional space. However, the definition of the curvature that we gave, following Gauss, used three-dimensional constructions—the tangent plane, the normal, the principal radii of curvature...[2]

But Gauss also found the way to describe the curvature in purely internal terms. He gave the means for our two-dimensional creatures, who stay on the surface and know nothing about the third dimension, to find out whether their world is flat or curved.

We consider two arbitrary families of curves on a surface. We repeat, the families can be quite arbitrary. Together the two families form a coordinate grid. Let $\vec{x} = (x_1, x_2)$ be the coordinates of a surface point. Now suppose we want to find the distance ΔS between two very close (otherwise, completely arbitrary) points $\vec{x}$ and $\vec{x} + \Delta\vec{x}$. The components of the vector $\Delta\vec{x}$ will be assumed to be very small.

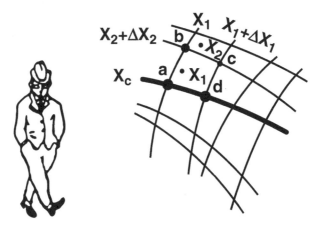

Gauss considered the following expression:

$$\Delta S^2 = g_{11}(\vec{x})(\Delta x_1)^2 + 2g_{12}(\vec{x})\Delta x_1 \Delta x_2 + g_{22}(\vec{x})(\Delta x_2)^2.$$

It is called the *basic metric form*. For a non-mathematician this formula is rather formidable in appearance. No need to fear, we will not use it. Only two remarks.[3]

---

[2] The latter *do* depend on embedding. It is only their product $R_1 R_2$ that does not.

[3] To clarify the meaning of this formula, one can think of the following question. The position on the earth surface is described by two geographic coordinates, the latitude θ and the longitude ϕ. Suppose we want to travel from the point (θ, ϕ) and the close (compared to the radius of the Earth R) point $(\theta + \Delta\theta, \phi + \Delta\phi)$ and would like to know what distance ΔS are we about to cover. A college student (and any seaman!) will

(1) The "physical" meaning of this expression is very simple. ΔS is the distance between the points $\vec{x}$, and $\vec{x} + \Delta \vec{x}$.

(2) The coefficients $g_{11}(\vec{x})$, $g_{12}(\vec{x})$, and $g_{22}(\vec{x})$ naturally vary from one point of the surface to another.

The important thing here is a result that Gauss obtained. He demonstrated that the curvature of a surface is completely defined by the functions $g_{11}(\vec{x})$, $g_{12}(\vec{x})$, $g_{22}(\vec{x})$. This is not all. He proved that no matter what system of coordinates is chosen, the curvature does not change. This is not self-evident in the least. Indeed, the functions $g_{11}(\vec{x})$, $g_{12}(\vec{x})$, $g_{22}(\vec{x})$ all change when we switch to a new coordinate grid. But the Gaussian curvature is built up out of them in such a way that it remains unchanged. In other words, the Gaussian curvature is completely independent of the manner of description.

It is an inner property of the surface. And so for two-dimensional surfaces the *essential* geometry (curvature) is determined solely by the basic metric form. This form depends on two variables. Knowing the coefficients, we could compute the Gaussian curvature of the surface at any point.

Now Riemann's idea can be conveyed in just two words. In a purely formal fashion let us examine similar expressions for three, four and n variables. We will say that such a metric forms define a geometry of a three-, four-, and n-dimensional world. We can compute (Riemann taught us how to do so) the Gaussian curvature[4] for such worlds. And then we will be able to say exactly what geometry will be accomplished in each one.

If the curvature [any one in the whole set of curvatures—A.S.] is different from zero, we say that the world is curved. And we would notice that even without going far from the place where we stay. All we have to know are the curvatures at that point.

The geometry of a "world" can be of any kind. At this juncture, it does not even matter very much what *is* this geometry. Riemann's theory provides for all conceivable cases.

That, roughly speaking, is all.

---

immediately tells us the answer:

$$\Delta S = \sqrt{R^2 \Delta \theta^2 + R^2 \sin^2 \theta \Delta \phi^2}\,.$$

It is nothing but a hypothenuse of the rectangular triangle with the sides representing the displacements along the meridian and the parallel.

And you see that this formula has the same form and the general formula in the text with $g_{\theta\theta} = R^2$, $g_{\phi\phi} = (R \sin \theta)^2$ and $g_{\theta\phi} = 0$.—A.S.

[4] And a number of other curvatures that appear in higher dimensions.—A.S.

It is simply a generalization of the Gaussian theory of surfaces to the case of many variables. At the beginning of the twentieth century, it turned out that it is precisely Riemann's geometry which we need to describe the actual world we live in. And not for three but for four dimensions, the fourth dimension being time.

We leave Riemann.

My task now is to keep calm and refrain from shouting hurrah.

Because throughout the whole of mathematics there are hardly a dozen ideas equal in sheer beauty to the proof of the non-contradictoriness of Lobachevskian geometry.

The whole structure rests on the fact that the mathematician cares not a whit about what lies behind his Basic Concepts—so long as the axioms are satisfied.

Up to a point, geometry is hardly more than a game in logic. The straight line, the point, the plane, motion are simply pieces used in the game. The only thing the mathematician knows about them is his axioms, the rules of the game involving the pieces.

At this stage, geometry is just as useless to the physicist as chess or dominoes. It is only when the physicist finds out experimentally that his quite real straight lines, points and so forth can be very precisely described by mathematical abstractions, only when he sees that the axioms of mathematics do indeed describe the behaviour of real lines, points, planes, etc., only then does geometry become one of the chapters of physics, the science which studies the world about us. Up to that point, geometry is a game of logic.

But it is just this logical formal approach that enables one to prove the noncontradictory nature of the geometry of Lobachevsky.

The scheme is the following.

There are two games: Euclid's geometry and Lobachevsky's geometry.

Let us attempt to demonstrate that if in the rules of one of them there is a hidden internal contradiction, then it will inevitably occur in the rules of the other one.

The rules of the game—I repeat—are the axioms.

You will see that we have somewhat changed the statement of the problem. We realize that it is a hopeless undertaking to attempt to prove rigorously the problem of noncontradictoriness for any single logical system.

No matter how many millions of theorems we prove, there will never be complete confidence that the next theorem will not contain a contradiction.

What we have set to prove instead is that if the geometry of Lobachevsky is contradictory, then the geometry of Euclid is unavoidably contradictory as well. One is not better than the other.

At first glance there is no clear way out here either.

The rules of the game (the axioms) in the two theories are different. True, they differ only in one axiom, that of parallel lines, but that does not change much. The games are different, and it is not clear at all how one can bridge the gulf between them.

Still and all, there is a way.

I am afraid that the variety of analogies brought in to illuminate the problem will only obscure it the more, and so I will start on the proof directly. The man who gave us this proof was one of the greatest mathematicians of the 19th century, Felix Klein. He was an interesting man, of great complexity, but unfortunately we cannot go too far into history. I wish to recall only one striking fact.

Klein lived a long life. If we take only the papers he wrote after the age of 30–35, he would be a magnificent versatile scientist by any standard. An active, subtle, fertile mathematician, and a brilliant expert in the history of his subject; he was one of the best teachers in the whole history of mathematics.

But once he made a harsh, categorical statement. He said that after the age of thirty, because of a nervous breakdown brought on by the investigation of a certain mathematical problem, he was never again capable of creative activity. This was not coquetry, either. It was exactly what he thought.

I like such people. It is quite a different question as to whether that makes their lives easier for them or not.

So here is the proof.

First we play Euclidean geometry. Consider an ordinary circle. Draw a chord. Take a point not lying on that chord. It is evident that one can draw through the point any number (an infinity) of other chords that will not intersect our chord. They will make up all the chords that lie between the two chords that intersect ours at the end-points (where it cuts the circle).

Clear so far. But what has this circle to do with the geometry of Lobachevsky?

Here is the miracle.

Klein's idea was to convert the trivial circle (or rather the interior of a circle—a disk without border) into a model of a Lobachevskian plane. The point is—we repeat—that the mathematician is quite indifferent to what his Basic Concepts refer. The ultimate thing is that his axioms be

satisfied. Now we can start playing the double game. We define a disk as a Lobachevskian plane, any chord in the disk as a Lobachevskian straight line, and a point as a Lobachevskian point.

Naturally, we also have to include in the game the properly defined Basic Relations: *to lie between, to contain* and *motion*. Add them. With these newly defined Basic Concepts at our disposal, we can play Lobachevskian geometry using the notions of Euclidean geometry.

To be able to do this, we have to check through our list of axioms and see whether our Concepts satisfy the axioms of Lobachevsky's geometry. It is comparatively easy to see that everything is in order with most of the axioms. Even—marvellously so, in fact—with the parallel axiom, which is the only one that distinguishes Lobachevsky's geometry from Euclid's. Indeed, one can now easily draw through a given point to a given "straight line" an infinity of "straight lines" that do not intersect it."

I give "straight line" in quotes, but once we prove that, for our Concepts, all the axioms of Lobachevsky's geometry are fulfilled, we will be able to remove the quotation marks.

Do not forget that we are playing a double game. All the time we have to translate from the language of Euclidean geometry into that of Lobachevskian geometry. And vice versa.

Everything is well with the notions *to contain* and *to lie between*. They remain the same in both languages. The difficulties begin when we go over to *motion*. The concept "motion" has to satisfy the entire group of axioms of motion. (Glance through them. They are at the end of Chapter 3.)

But we want to interpret our disk as the Lobachevskian plane. It is clear then that its "motion" cannot coincide with habitual Euclidean motion, it is something different. Indeed, motion, it will be recalled, is a one-to-one mapping (transformation) of a plane onto itself. This means that, in Euclidean language, a motion is a transformation of Klein's disk onto itself. But ordinary Euclidean motions that include translations do not fulfil this requirement. A translation displaces the disk to some other position on the Euclidean plane.

Still it *is* possible to formulate the concept of motion of a non-Euclidean plane in the language of Euclidean geometry.

One class of transformations insistently claims our attention. These are simple rotations of the disk about its centre. In contrast to translations, they leave the disk at its place. However, it is easy to see that these transformations cannot alone be used as candidates for "non-Euclidean motions".

In rotations, it is not possible to transfer any given point of the disk to any other preassigned point. For example, the centre of the disk. In such transformations, it is always a fixed point, passing to itself. On the other hand, the axioms defining motion require that any given point can be transferred by a certain motion to a different point. Thus, rotations do not satisfy us.

Yet the transformations of the disk that we need exist. That is the central and most radiant part of the Klein scheme. He pointed out an infinite number of such transformations of the disk (they are called projective transformations) which transfer a disk into a precisely identical new disk, with any internal point of the old disk passing to an internal point of the new disk. Any point of the circular border of the old disk remains on the circular border of the new disk. And the chords of the old disk pass to chords of the new disk.

These transformations of the disk (projective transformations, in Euclidean language) satisfy, in non-Euclidean language, all the axioms of motion.

For instance, the fact that chords go to chords signifies in non-Euclidean language that straight lines pass to straight lines, etc. Now comes the last and decisive step. We refer to these transformations as *motions of a Lobachevskian plane*.

We can summarize the foregoing in the following table.

KLEIN MODEL

In the language of Euclid's geometry	In the language of Lobachevsky's geometry
Disk	Entire plane
Chord	Straight line
Point	Point
To contain	To contain
To lie between	To lie between
Projective transformation of the disk onto itself	Motion

All the properties of projective transformations are of course known, but we do not need to learn them. All we have to do is accept the fact that such transformations exist.

And so—this is the minute we have been waiting for—if it is possible to declare the disk a Lobachevskian plane (and we have proved this), then the problem is solved.

Indeed, suppose that in proving some kind of theorem in the geometry of Lobachevsky, we arrive at a contradiction.   But every theorem of Lobachevsky's geometry is now also some theorem of the geometry of Euclid.

Each theorem may be stated in two languages. If we have a contradiction in Lobachevsky's geometry, we also, at the same time, get one in Euclidean geometry.

Of course, in Euclidean language the contradiction will look differently and will open up in a different theorem, but that is quite immaterial. The important thing is that if one of the geometries involves logical contradictions, this will be the case for the other geometry as well.

The geometries are equivalent.

This then proves the independence of the fifth postulate of all the remaining axioms of Euclid's geometry.

That is all!

But in science, like in the Arabian Nights, the end of one story is but the beginning of the next.

Proof of the noncontradictoriness of the geometry of Lobachevsky signified for mathematicians the start of a colossal cycle of studies in axiomatics, the creation of a highly intricate, ideally rigorous and absolutely abstract apparatus of mathematical logic, an apparatus that was infinitely removed from the slightest practical application—until it was found that computers... Well, I think that we should not pursue this discussion.

Better let us return to the Klein model to note a very amusing point. Take two points within our disk. Draw a chord through them. In the language of Euclid the distance between these points is equal to the length of the chord. But what is the situation in the language of non-Euclidean geometry?

Intuitively, we can see that at any rate it cannot be equal to this length. Indeed, distances between two points on the infinite Lobachevskian plane can be arbitrarily great, while the "Euclidean distances" between points of our disk are restricted by its diameter. It is clear that we have to define a "non-Euclidean distance" in some other way. But how? Very simply if we recall how the concept of length is introduced into geometry.

Roughly, it is done as follows. Take a scale unit—some segment—and by means of transformation of motion make it coincide with the segment being measured. The length of the latter is determined by the number of times the scale unit fits into it. We will not go any further. The important thing to note is that the definition of equality of line segments (and consequently of their lengths) as also, incidentally, the congruence of any geometrical figures, is determined by means of the concept of motion.

That is the situation both in the geometry of Euclid and in the geometry of Lobachevsky. But in our model, motion in the Lobachevskian plane is, in Euclidean language, a projective transformation of a disk. Therefore, it comes out that in the language of Lobachevskian geometry, two line segments are equal if one maps to the other in a projective transformation. Recalling that the length should not change during transformation of motion, we realize that the "non-Euclidean length" must remain the same in a projective transformation. It must, as mathematicians say, be invariant under a transformation. This quantity—the invariant—is known, of course, for projective transformations of a disk. If we also take into account that the length of the sum of two line segments must be equal to the sum of the lengths of these segments, it turns out that the "non-Euclidean distance" is determined uniquely. And, of course, such a distance behaves as expected (that is, it becomes infinite) when one of the points lies on the circular border of the disk.

The circular border of the disk corresponds to infinitely distant points of the Lobachevskian plane.

A somewhat extravagant character of "non-Euclidean motion" in the Klein model is also evident in the fact that the size of a "non-Euclidean angle" between two straight lines is quite different from the angle between two chords in the Euclidean language. But these are only details. Important details, but details. All the essentials have already been stated.

And now the last point.

To prove the noncontradictory character of the solid geometry of Lobachevsky, it suffices to convert the Klein disk into a ball.

A few years after Klein, the French mathematician Poincaré proposed another model of Lobachevskian geometry. Also in the finite ball. It is perhaps even more remarkable. Poincaré even thought up a marvellous world of physical beings, which, from the Euclidean viewpoint, would live in the finite ball of Poincaré, but from their own point would claim that they were living in the infinite space of Lobachevsky.

In this world, the straight lines of Lobachevsky are, in Euclidean language, arcs of circles perpendicular to the surface of the bounding sphere. The accompanying drawing will give the reader some idea of Poincaré's model. There are a lot of attractive features in the Poincaré ball, but we will have to call a halt at this point, for other things claim our attention.

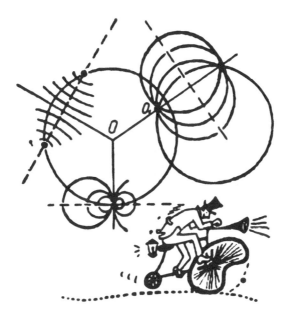

Chapter 11

An Unexpected Finale.
The General Theory of Relativity

We are in a predicament.

Up to this point, we had been mostly talking the language of high school.

Well, in some parts of the preceding chapter we exceeded that level, but, hopefully, we were still able to convey to some extent the essence of the proof of noncontradictoriness of the geometry of Lobachevsky and to impart to the reader some ideas of Riemann. Though it was not easy for the author and, probably, for the reader.

But things have now become more complicated. To understand more or less well the content of the general theory of relativity, one has to learn the special theory first. But the author does not have the right to require such knowledge from his readers and does not have a possibility to elucidate it on the pages of this book.

The most honest way out would be to say nothing—simply not to write this chapter. The temptation was great. But that would mean playing the whole symphony of the fifth postulate without its triumphant, purely Beethoven finale.

That would be a pity to do so, and I've decided to go ahead no matter what. All that I can do is to warn you that what follows is only a bare outline, extremely superficial.

The general theory of relativity is based directly on the idea of the "non-Euclidicity" of space. That is what interests us most. And so let us try to dispense with the details of special theory of relativity. We will confine ourselves to only a word or two.

Geometry after the year 1905

The special theory of relativity has already substantially altered our views concerning geometry. To begin with, let us try to grasp the connection between geometry and physics in general and also to see what has changed in geometry as a result of the special theory of relativity.

Before Einstein, the universal and firm conviction was that Euclidean geometry reigns supreme in the real universe in which we live. There were no reasons to think otherwise. The theoretical possibility that our world is describable by some kind of non-Euclidean geometry remained a purely theoretical one, while Gauss', Lobachevsky's, and Riemann's suspicions on this score were no more than speculations. The situation was as if someone said: "The supposition that so-and-so, Mr. X, is a Martian dweller does not in the least contradict the laws of formal logic."

"That may be," would be the response, "but all observations and experiments point to so-and-so being an inhabitant of the earth."

Now after the advent of the special theory, there appeared real doubts about the problem of the origin of Mr. X being so crystal clear.

Throughout the book we discussed geometric problems staying in the mathematicians' camp. We must move now into the camp of physicists.

Let us see what geometry means to mathematicians and physicists.

To the mathematician, geometry, as we have said time and again, is essentially a formal game with the Basic Concepts and axioms chosen for them. He requires the game to obey the rules of formal logic, and he does not care so much whether his geometry can aspire to any relationship with the actual world in which we live.

True, every person was unequivocally convinced that it is Euclidean geometry that reflects the properties of our universe. But that was simply taken for granted. A sort of natural trait of the human mind. The fact that geometry has an experimental foundation was somehow forgotten. What is more, for two thousand years prior to Lobachevsky, geometry was carefully guarded against the defiling effects of experiment; it was kept away from any kind of "empirical basis".

Einstein rather maliciously but precisely remarked that what happened to the axioms and Basic Concepts was similar to the process of converting the heroes of antiquity into gods. In place of a realistic basis, there arose the "myth of geometry", a rather hazy conception of axioms as something "intrinsic to the human mind, to intuition, and to the spirit". It is hard to grasp the meaning of the last words, possibly because there isn't any.

However, it must be said that the hypnosis of abstraction was so great that it held the greatest minds spellbound. Physicists were among them. One can even mention some outstanding names, people not without talent— Isaac Newton, to take one instance.

His Basic Concepts that are given in the opening chapter of the *Principia* are fundamentally unobservable and unknowable. Newton's *absolute space* and *absolute time* are something "intrinsic to the human (and perhaps also divine) consciousness". There is no irony here, none in the least. It precisely conveys the substance of the notions "absolute space" and "absolute time".

So physicists too were engaged in "turning heroes into gods". If we continue the divine analogy, it will be seen, however, that because of their scatter-brainedness, physicists, though theoretically recognizing and preaching the religion of the absolute, actually paid no attention to it, and did not draw any real conclusions therefrom.

The first example was set by Newton himself.

He formulated all the laws of his mechanics for "absolutes", but employed them in the solution of quite concrete problems. Since, essentially, the axiomatics did not interfere in any way, no attention was paid to it.

In this sense, mathematicians turned out to be more consistent in their attitude. They had already fully analysed the problem of axiomatics when physicists were just beginning to take a serious interest in the foundations of their science, the basis of their conceptions concerning space and time.

On the other hand, though, the physicists advanced much farther and at one step. Here almost all the credit goes to one man—Einstein.

It was about this time that the attitude of physicists to geometry became clear-cut. Intuitively, subconsciously they always believed that the entire problem of interrelationships between geometry and physics was rather artificial.

Now the situation was substantiated with complete rigour. The point was this. The Basic Concepts of geometry are abstractions of our conceptions of actual physical objects. For example, Einstein says that rigid bodies with markings on them, realize (given due caution) the geometric concept of a line segment, and rays of light realize straight lines. He then goes on to say that, if one does not adhere to this viewpoint in practice, it is impossible to approach the theory of relativity.

But if that is the case, then geometry is simply a chapter of physics! Its first chapter!

Practically speaking, what we have just said does not change matters much. We have dethroned the axioms and Basic Concepts, we have reduced geometry to a generalization of physical experiments and now see that the truth or falsity of geometry is a question of experiment, but all the specific assertions have remained unchanged.

We recall that, essentially, Gauss and Lobachevsky and Riemann all thought similarly. They defended the positions of the practical physicist.

However, if we consistently develop our views, it will be seen that we have already proved a few things. Things that are new and important. What is more, our views suddenly lead us to certain doubts as to the actual realizability of geometry. Here the attack is from fresh positions.

One of the principal chapters of any geometry is that of the geometric theory of measurement. In order to develop geometry, we have to define the concept of length with full mathematical rigour. This was naturally done by geometers. Their definition of length lays on two "whales".

We need:

(1) A line segment whose length is taken to be unity.
(2) A procedure for measuring, which in geometry amounts, roughly speaking, to laying off the scale unit on the segment being measured and counting the number of times it is taken. The resulting fractional number of times (it may accidentally be integral) is the length of the line segment.

In that way, one can, say, measure the length of a side of a triangle. In doing so, we tacitly assume that if the triangle is at rest relative to the "scale unit" or is in motion the result will be the same. Now since we said that all geometrical objects are an idealization of actual physical bodies, then the words given above cease to appear so clear.

If the triangle being measured is in motion relative to the scale unit, our procedure for measuring is no good at all. If we stand on the platform of a railway station and wish to measure the length of the doors of a railway coach of a train passing by at high speed, we cannot apply the scale unit— the coach will not wait for us. To perform the measurement, we would have to be moving along in the same direction and at the same speed as the train (with the scale unit in our hand). But then both "unit" and "object being measured" would be at rest relative to one another, and we return to the situation considered above.

Obviously, some kind of new procedure is needed for measuring moving bodies. But if our procedure is new (it matters little what kind, so long as it is new), then we are not positive in any way that our new "length" will coincide with the earlier one.

Actually, we have introduced a totally new concept. From the standpoint of formal logic there are no grounds to expect that it will coincide with the earlier one. Only experiment can resolve the matter.

Let us stop for a moment.

A little thinking will make it clear that these are very unpleasant words for the axioms of geometry.

We assert that our geometrical concepts, generally speaking, can change if the actual solids whose geometric properties are under study are moving relative to us.

We say, "something can change in this case". We thus demand that the geometric system of axioms be supplemented by fresh axioms of a purely physical nature.

Consistently developing our views, we become convinced that there should be a rather large number of such axioms. Indeed, all our segments (including of course the scale unit) are abstractions of actual solids. But, as we know, solids expand when heated, their length changes. Measurements with cold and hot scale units will give different results.

Consequently, if we want to be absolutely precise (and that is our aim, isn't it?), we must introduce into geometry the notion of temperature and fix a constant "temperature of the scale unit".

However, temperature is not the only thing that affects physical properties. Hence, we will have to specify all the physical conditions. It then works out that only if all manner of precautions are taken can we hope that the axioms of "pure geometry" will describe our universe correctly.

And the latter is the only thing that now worries us.

Generally speaking, this work has not yet been done in all its details. Probably it is not very much needed, though possibly it is. At least twice it has turned out that redefining the physical conditions in which the geometry of the world was constructed has completely overhauled our conceptions of nature.

The first time this occurred when the special theory of relativity was introduced. It was found that the length of a moving segment differs from that of a segment at rest.

We will not go into how all that came about and will confine ourselves to a few general remarks.

1. We are not abashed by the fact that the length of a moving segment comes out different from that of a segment at rest. We realize that determining the length of a moving body involves a new procedure of measurement, and hence, strictly speaking, it is a new notion. It need not coincide with the old notion.

We also realize that this new notion must in some way or other be introduced, for we are no longer playing a game of logic but are creating a tool with which to study the actual world. Our Concepts must be able to describe this world fully and well. That is what they are for.

They appear as a result of the study of the real physical world. But there are moving bodies in the world and one has to be able to describe them.

2. It turned out that without employing the concept of time it is impossible to determine, logically and well, the "length of a moving body".

We become suspicious at this point.

It is disconcerting that a new and highly important concept—Time—enters into our geometry. Up to now geometry had been associated solely with Space.

But—we continue to reason—maybe there is no need to worry. If the length of the moving segment coincides exactly with the length of the segment at rest, there will be no essential change.

In this case, the notion of Time will in no way be associated with that of Space.

Now if experiment shows that the length of a moving segment is different, and if it turns out that this length depends on its velocity, for example if it diminishes in accordance with the law

$$l_{\text{moving}} = l_{\text{rest}} \sqrt{1 - \frac{v^2}{c^2}},$$

where v is the velocity of the moving segment and c is the velocity of light... If the velocity and, via the velocity, the time too enter geometry... then we will have to say: time and space are interrelated. Then in geometry it will be impossible to study space independently of time.

And that is exactly what Einstein demonstrated! All our expressive "ifs" come true.

The length of a moving body is indeed dependent on the velocity: *time enters geometry, the properties of time turn out to be dependent upon the properties of space*, and all our earlier views concerning the universe and geometry prove to be only a rather naive approximation. It is only when we confine ourselves to studying the cases when the relative velocities of objects are small that our old conceptions function properly and we can regard space as being independent of time, and time as independent of space.

In this case, the old good geometry of Euclid is a proper instrument for studying space. Then we can take it that the properties of space do not depend on time.

Such were the ideas that arose in the year 1905. They constitute the content of the special theory of relativity.

The inner logic and elegance of Einstein's theory were so striking that within three or four years all the leading theoretical physicists became its enthusiastic adherents. In 1909 Max Planck exclaimed: "It need hardly be said that the new—Einsteinian—approach to the notion of time demands of the physicist an ultimate capability of abstraction and an enormous capacity for imagination.

"In its audacity, this theory surpasses everything achieved up to this time....

"Non-Euclidean geometry, by comparison, is child's play. Yet, in contrast to non-Euclidean geometry, whose application can seriously he considered only in pure mathematics, the principle of relativity has every right to pretend to a real physical significance.

"In its depth and consequences, the upheaval wrought by the relativity principle... may be compared only with that effected by Copernicus."

Planck was right but he did not know that that was only the beginning.

To summarize: a new concept of four-dimensional space-time entered physics together with the special theory of relativity. But, as before, three-dimensional space is described by the geometry of Euclid.

True, in that same year of 1909 a very curious fact came to light. It was found that the law of composition of speeds in the special theory of relativity coincides exactly with the law of composition of vectors in the space of Lobachevsky.

In other words, the formal space of relativistic velocities is Lobachevskian space. But this appeared to be a purely formal coincidence. Neither at that time nor later was any profound physical meaning found in this analogy.

Still more sensational and startling news followed.

Physics and geometry (after 1916)

Planck was not to blame, because if one needs an instance of the most unexpected discovery in the history of science, then this is the general theory of relativity.

For three hundred years the foundations of the theory of gravitation were in a state of absolute rest. Newton gave the law. And that was all. Actually, there was just one fundamental formula that lay at the core of calculations of the motion of celestial bodies in all the numberless volumes of subtle, elegant and magnificent investigations into celestial mechanics:

$$F = \gamma \frac{m_1 m_2}{r^2}.$$

This states that the force of attraction of any two bodies in the universe is proportional to the product of their masses and inversely proportional to the square of the distance between them.

The quantity γ is a dimensionful constant: $\gamma = 6.67 \cdot 10^{-11} \, N \cdot m^2 \cdot kg^{-2}$.

The law of universal gravitation is truly magnificent! One could go on singing the praises of the "simplicity" of Newton's ideas, but that is a waste of good time. The simplicity lies only in the analytical form of the law. This "naive" formula summarizes several not-so-obvious, subtle and—what is more—at first glance strange physical assumptions. It required a Newton to produce it. Over one hundred years passed before the law of gravitation was unconditionally accepted. Note too that the protests came not from ignorant people or obscurantist scholars, but from the greatest and most talented scientists of the day. And so the talk of simplicity can refer only to the magnificent harmony of nature, to the beauty and elegance of her basic laws.

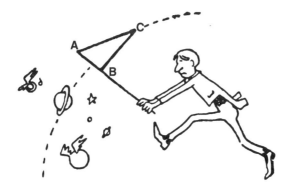

Newton told us how gravitation operates. But no word was said about why it functions in precisely the way it does.

By the beginning of the twentieth century, people had almost reconciled themselves to this situation. It was as if looking at the smoothly polished surface of your furniture you find it difficult to imagine the rough unworked wood that lies underneath.

Incidentally, attempts were made from time to time to offer some mechanism for the law of gravitation, but they all invariably came to nought. When science was rapidly developing, physicists had their hands full of specific urgent problems, and during periods of decline and quiescence there was neither the enthusiasm nor the moral energy to risk investigating such a cardinal and most certainly hopeless problem.

If for the beginning a Newton was necessary, then for the continuation an intellect of perhaps a still greater scale was needed.

Most likely, one must agree with Einstein that without him the theory of gravitation might not have been created to this day.[1]

In science (in the arts too, by the way) the role of a genius is perhaps greater than in other fields. One man is capable of accomplishing more than hundreds of huge research teams. The decisive factor is not quantity but quality.

So between the years of 1905 and 1915 Einstein studied the problem of gravitation. By the end of 1915 the work was completed. During this same period he was engaged in many other things and, in passing as it were, he obtained fundamental results in solid-state theory. But all the time uppermost in his mind was the general theory of relativity. It continued to occupy this central place to the end of his days.

Of course, before going to the heart of the matter, we will, as we have been doing all along, start out with a few general ideas and a story or two. When one is dealing with Einstein and his works this is all the more necessary, because... well, you will see it soon.

Once, reading a hunter's journal—I can't even imagine how that ever got into my hands—I came upon an article about snakes. I started to read with an interest, though probably would not go beyond the title if the subject were gophers rather than snakes. The author, who had captured about 1,500 snakes, reported, among other things, that not one of these snakes had ever attacked him first.

This was amazing. I read to the end. The article was a serious one

---

[1]Maybe an exaggeration.—A.S.

written by a professional snake-catcher. He analysed a variety of specialized problems, stressed the value of venom, criticized the situation in the country in that respect, and, what was particularly interesting, one felt that he liked all these poisonous snakes and considered them very useful.

The problem of boosting the venom output of a Central Asian cobra was discussed as if one were talking about Kholmogory cows. The article began with the statement that sensational stories about snakes do more harm than good. And he listed a few of the mistakes that journalists make. I gathered that he was truly upset and that he very much wanted people to get a clear picture of this complicated and rather tedious profession of snake-catcher, in place of all sorts of "romantic horrors".

Now, I recalled this story not only because I wanted to amuse my readers but as an illustration that people often get far-fetched, highly distorted conceptions about things with which they do not personally come into contact.

And, unfortunately, the conceptions that most people have about the pecularities of the profession of a scientist (especially that of the physicist) is at about the same level as their conceptions about the work of a snake-catcher.

More than anything else, the theory of relativity (and of course Einstein himself) suffered from sensational stories.

It was his luck he could dismiss with calm and indifferent irony the endless uproar around his name that continued from 1919 onwards. One can perhaps only offer prayers of gratitude that all the publicity had practically no effect on his good nature.

But so much nonsense was whipped up around the theory of relativity, both general and special, that one feels embarrassed. True, physicists themselves are somewhat to blame too. For many years, even in professional circles, it was believed (and still is perhaps) that the ideas of relativity theory are very complicated. Particularly if one is dealing with the general theory.

This was quite natural during the first years after Einstein's work appeared. It is always the case. From what you have seen in this book, I hope it is clear that even such an elementary (if judged without prejudice) idea as that of Lobachevsky was first grasped only with exceptional strain.

But fifty years have now passed since the creation of the general theory and sixty years since that of the special theory of relativity. We should have long since put everything in its place and realized that the fundamentals of Newton's mechanics are, at any rate, hazier and, possibly, more involved than the principles of the theory of relativity.

Even from the most general reasoning it is clear that it could not be otherwise. In both cases we deal with the same things—the fundamental ideas of space and time. And the farther we penetrate into the essence of the matter, the clearer, simpler and more harmonious do our conceptions become.

In building his general theory, Einstein proceeded, as he himself said, from a childish, naive question that had engaged him ever since his school days.

"What happens in a falling lift?"

Another ten years of intensive work, several dozen faulty versions that had promised success, and a number of probing investigations were needed before the problem was resolved in 1915.

Crudely speaking, that was how things actually stood.

From outside, it looked differently. An excerpt from Chaplin's autobiography gives us a picture of how all this appeared in the minds of two non-experts, who cannot be suspected, however, of the slightest desire to twist the truth.

"As Mrs. Einstein had requested it should be a small affair, I invited only two other friends. At dinner she told me the story of the morning he conceived the theory of relativity.

"The Doctor came down in his dressing gown as usual for breakfast but he hardly touched a thing. I thought something was wrong, so I asked what was troubling him. "Darling", he said, "I have a wonderful idea." And after drinking his coffee, he went to the piano and started playing. Now and again he would stop, making a few notes then repeat: "I've got a wonderful idea, a marvellous idea."

"I said: "Then for goodness' sake tell me what it is, don't keep me in suspense."

"He said: "It's difficult, I still have to work it out."

"She told me he continued playing the piano and making notes for about half an hour, then went upstairs to his study, telling her that he did not wish to be disturbed, and remained there for two weeks. Each day I sent him up his meals, she said, and in the evening he would walk a little for exercise, then return to his work again.

"Eventually, she said, he came down from his study looking very pale. "That's it," he told me, wearily putting two sheets of paper on the table. And that was his theory of relativity."

Most likely something very much like what is described here actually took place. It might be literally true. Mr. Chaplin of course wrote the way he saw things. But this changes nothing at all. If that is the truth, then it is only a minute particle of the truth.

Without probably realizing it himself, Chaplin perceived the story that Mrs. Einstein told him as a film-maker. And his own narration resembles a sketch of a dramatic and interesting, but not very deep screenplay.

Now I am about to undertake what I myself have so harshly criticized: a very superficial and therefore unavoidably distorted description of the general theory of relativity and its interrelationships with geometry.

Einstein had two guiding ideas. One did not seem, at first glance, to have any relation whatsoever to geometry. That was the lift. Or, to put it differently, the assertion of the equality of an inertial mass and a gravitational mass. That was the one and only experimental fact upon which the entire theory was constructed.

There is nothing more amazing in the whole history of science.

Let us explain what the inertial mass and gravitational mass mean. Everyone should know Newton's second law. However, I suspect that not all our readers do have a full grasp of either that law or of the other laws and, in general, of the fundamentals of classical mechanics. Unfortunately, school physics only performs a few formal manipulations with Newton's laws and does not demand much understanding on the part of the student.

Yet—and I am prepared to repeat this without end—to grasp

thoroughly the fundamentals of classical physics is tantamount to fully preparing oneself for an understanding of the theory of relativity, because as soon as the notions of space, time, force and mass cease to exist as nebulous and purely intuitively perceived entities, as soon as their exact meanings have been elucidated, then any physical theory will appear as a consequence of a definite system of axioms. Now, a choice of axioms is determined by experiment.

I admit that this is my sore spot, and since we haven't space enough to give a clear analysis of the basic notions of physics, my suggestion is that the reader consult a book or two on the subject.

For the present, suppose that the reader is familiar with Newton's second law and even has fully mastered it.

The proportionality constant between a force and the acceleration is the mass m that determines the inertness of the given body. We shall call it the inertial mass m_{inert}.

Newton's law of universal gravitation refers to the gravitational interaction of bodies.

A priori, there are absolutely no grounds to believe, not the slightest hint, that the formula which determines the force of interaction must somehow be dependent on the inertial mass. For classical physics, this is a still more unexpected and inexplicable fact than, say, the dependence of the number of weddings in Vladivostok on the weather in the Antarctic. In the latter case, we at least have a logical link-up in that the Soviet whaling fleet is based at Vladivostok. Now, in the case of the gravitational and the inertial masses there was no clarity up to the time of Einstein.

There was a remarkable experimental fact, and everyone, Newton first, made note of the marvellous coincidence. Many experiments were carried out over the years up to the beginning of the twentieth century. The last experiments—those of Lorand Eötvös—were amazingly accurate. The idea behind all the experiments was extremely simple and we shall now examine it. First we will write down again the law of gravitation.

We will write the masses as m_{heavy}, for we do not know whether these masses are the same as m_{inert}. We want to find an experiment where this assertion can be checked. So we have

$$F = \gamma \frac{m_{1\,\text{heavy}}\, m_{2\,\text{heavy}}}{r^2}.$$

Let us examine the concrete case of a freely-falling body. The force compelling it to fall (the force of gravitational interaction) is the force of gravity.

On the other hand, if we know the acceleration and the inertial mass of the falling body, say a small ball, we can find the force by means of Newton's second law. We thus have two equations:

$$F = \gamma \frac{m_{\text{heavy}} M_{\text{heavy}}}{r^2}, \qquad (1)$$

where M_{heavy} is the gravitational mass of the earth, and r^2 is the distance from our ball to the centre of the earth. (Newton established that a massive sphere attracts with the same force as if its entire mass were concentrated in the centre. That was a purely mathematical problem.)

$$F = m_{\text{heavy}}\, g, \qquad (2)$$

where g is the acceleration of free fall. Combining the two equations we get

$$g \frac{m_{\text{inert}}}{m_{\text{heavy}}} = \gamma \frac{M_{\text{heavy}}}{r^2}.$$

Now if $m_{\text{inert}} = m_{\text{heavy}}$ for all conceivable bodies; if they are equal in the case of steel, wood, gases, liquids and radioactive elements and polymers and so on and on, then $g = \gamma M/r^2$.

In other words, the acceleration of the earth's gravity is the same for all bodies.

This was first established by Galilei. And then, as we have already noted, the equality of the inertial and gravitational masses had been firmly established in dozens of experiments.

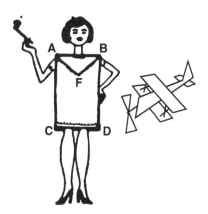

With the advent of the special theory, when it became clear that every kind of energy possesses an inertial mass, special experiments were performed with radioactive substances.

It turned out that in their case too the inert and gravitational masses were equal. That is to say, energy possesses heavy mass as well, which is the same as the inertial mass. In short, precise experiments demonstrated the identical equivalence of the inertial mass and the heavy mass. However it was one thing to know and quite another to understand. Einstein wanted to find out *why* they are equal.

It may not yet be clear what all this has to do with geometry, but nevertheless this sole experimental fact plus the special theory of relativity, plus one more requirement of a purely theoretical character was enough for Einstein to formulate his general relativity that brought about a complete change in our conceptions of the geometry of the universe.

Now about the theoretical requirement. We can even formulate it in strictly technical language: *the laws of nature must be generally covariant,* or, more simply, *all reference systems must be equivalent.*

I fully realize that this is not much of an explanation, I gave the statements more for my own consolation. We simply do not have the necessary time to go into the origin of the general theory of relativity. And I do not wish to give only a semblance of an explanation, though that would be fairly easy to do. The only thing I ask you to take on trust is that the "equivalence of reference systems" is a demand which largely stems from aesthetics. The inner logic and the beauty of a physical theory were to Einstein one of the most decisive factors.

It may be that he occasionally overestimated the relative significance of such arguments, but he believed that the laws of the universe should in principle be very natural and logical, and that theoreticians often distort them perceiving things in a crooked mirror. One can, of course, find fault with such reasoning—no things exist that do not have weak spots—but the fact that for him this mode of reasoning was efficient is proved by the results he achieved.

"The theory of gravitational fields constructed on the basis of the theory of relativity bears the name of the general theory of relativity. It was created by Einstein (and formulated in final form by him in 1916) and is perhaps the most beautiful of existing theories. The remarkable thing is that it was constructed by Einstein in a purely deductive fashion and only subsequently corroborated via astronomical observations."

Thus wrote Landau and Lifshitz in their fundamental course of theoretical physics which is considered to be the best in the world. And it is the only place in all their ten volumes where the authors display any emotion.

That fact alone speaks volumes.

But let us get back to apocrypha.

In reply to the query of his nine-year-old son, "Papa, what is it that makes you so famous?" Einstein is reported to have said quite seriously that when a blind bug crawls over the surface of a ball, it does not notice that the path traversed is curved. Said Einstein, "I, on the contrary, had the good fortune to notice that."

This passage is often quoted, but don't think that it exhausts the content of the general theory.

But it is obvious that Einstein himself believed that the basic result of his work was a fundamental change in our conceptions of the geometry of the universe.

We have already said that the special theory killed the idea of the geometric properties of space being independent of time.

Time had become a part of geometry.

But the properties of time only affected the geometry of moving bodies. For bodies at rest, the geometry of Euclid held true.

A new physical factor appeared in the general theory of relativity that determined the geometry.

The old result—the mutual dependence of the properties of space and time—was naturally retained. But this was not all. It turned out that the geometrical properties of the world at a given point at a given instant of time are determined by the gravitational field at that point.

This last phrase probably does not mean very much to the reader, and so we shall give a few precise statements and then a crude analogy that should demonstrate certain things.

In the general theory of relativity, the world is described by the geometry of Riemann. Here, when speaking of the world and its geometry, we all the time have in view the four-dimensional world. Time is inextricably woven into the geometrical properties of space.

As you recall, both Gauss and Riemann regarded the curvature of space at a given point as the determining characteristic. Also decisive were another "intrinsic characteristic of space"—the properties of the shortest lines (geodesics). These lines are physically determined by the trajectories of rays of light (and, more generally, of material particles not subject to the action of nongravitational forces; we have seen that such a trajectory does not depend on the physical nature of a particle).

According to Einstein, both the curvature at a given point and the properties of geodesics are determined by the gravitational field. Or, better to say, they *are* the gravitational field. In the general theory of relativity, gravitation occupies an exceptional place of honour. Roughly speaking, it is the most important of all interactions. It determines the geometry of the universe.

On the other hand, gravitation depends on the distribution of masses in the Universe. The result is that the geometrical properties of the world are determined by the distribution of gravitating masses.

We repeat again: whenever we speak of geometrical properties, we have in view a four-dimensional world, so that in ordinary parlance one ought to say:

The geometrical properties and the properties of time are completely determined by the distribution of masses in the universe and their motion.

And just like the geometry of the plane is approximately fulfilled for small areas of a two-dimensional curved surface, so small regions of the four-dimensional world may be approximately regarded as regions in which the curvature is zero.

Physically, this means that in small spatio-temporal regions one can exclude the gravitational field (an observer in a free-fall lift does not feel it; he is in a weightless condition) and pass over to the special theory of relativity.

According to Einstein, geometrical properties of space and time show up only when there are material bodies in the universe.

That, very roughly speaking, is the gist of the ideas of the general theory of relativity.

Two remarkable circumstances stand out in the story of the development of this theory.

1. At first Einstein was not even acquainted with Riemann's ideas. He had wanted to explain the equality of the inertial and heavy masses and found in his search that Riemann's geometry was the necessary mathematical form for a description of his purely physical reasoning.

2. The general theory is probably the only instance of a physical theory created in a purely deductive manner. There was only one experimental fact underlying the whole theory.

Today, the general theory has been corroborated experimentally a number of times and was just recently verified under laboratory conditions.[2]

Now the analogy which I promised.

Imagine a piece of cloth stretched taut. This is a plane. The geodesics

---

[2]The author meant here one of the predictions of general relativity—the *gravitational redshift*, that is the change of photon's wavelength when it goes up or down in the gravitational field. This prediction was confirmed in 1959. Quite recently, in 2016, another spectacular confirmation came—the direct observation of the *gravitational waves*.—A.S.

on it are straight lines. The curvature is zero. A free material particle on such a surface will move in a straight line. This is an analogue of the space-time of the special theory of relativity. Now throw a stone into the middle. The cloth sinks in the vicinity of impact. The shape will be distorted. The geodesics will no longer be straight lines. A particle in motion on such a surface, even in the absence of forces, will be deflected from a straight-line path.

The farther away from the stone, the less the curvature, and at infinity, the cloth is again flat. The whole cloth is a rough model of space-time in the presence of a pointlike gravitating mass.

And now the last question. What is the actual geometry of the world we live in?

Experiment has shown that at least in our part of the universe the curvature of space-time is positive; crudely speaking, that is, because the question of the true geometry of the universe is a very touchy one. Physicists have to use their imagination. This is a realm where hypotheses abound.

Formally speaking, the whole problem consists solely in determining the coefficients in the formula that defines the square of the distance in a four-dimensional world: space plus time. That is all!

As of today we have a number of models of the world. Several hypothetical universes. But we still do not know for sure which one fits the world in which we live. The portion of the universe accessible to the most powerful telescopes (only a paltry ten thousand million light years) is far too small.

Of course the local geometry of space-time varies from point to point and changes very fancifully near gravitational masses.

Let's try another analogy. Compare our situation with a dweller of a mountainous region of the earth attempting, with the aid of geodetic observations, to establish that the earth surface is a sphere. His region of observations is of course very restricted. Only a few kilometres. Quite obviously such a land-surveyor would not be in easy position.

Even if he is able—using his measurements—to find that the mean radius of curvature of his portion of the surface is 6,400 kilometres (the approximate radius of the earth), he will not be one hundred percent confident that the surface of the planet has the same curvature in regions outside his view. And then he will inevitably have to do what Isaac Newton so disliked. He will have to frame hypotheses.

That is the actual situation of down-to-earth physicists when they are asked about the geometry of the world in the large.[3]

Let's stop for a moment at this exciting point. It is time to do some summing up.

The creation of non-Euclidean geometry paved the way to two major scientific developments.

The first is the creation of axiomatics and, subsequently, of mathematical logic. This was initiated by Hilbert. We have already had occasion to mention him, but our story was very crude and approximate. Particularly, as regards the problem of the completeness of the axioms. I could have done a better job, but only at the expense of drawing out our story.

Anyway, while writing this book I did not find a way to tell the story of axiomatics precisely, concisely and comprehensibly. Too little has been said of axiomatics, and most of that was not very accurate. The only thing that I can do is to *advertise* some ideas that we have not touched upon in the main text.

The whole range of problems involving axiomatics is amazingly elegant. Even the way that many of them are posed is sometimes totally unexpected. This is particularly true of the problem of completeness. One result will suffice as an illustration. Already in the nineteen thirties the following theorem had been proved.

Suppose you have a certain logical system. Its foundation consists of the

---

[3]Now we know it somewhat better than people knew back in 1965 when Voldemar Smilga wrote this book. The global spatial curvature of our Universe is slightly negative, but the main point is that the Universe is not static—it expands and expands with some acceleration.—A.S.

Basic Concepts and the axioms. Say, Euclidean geometry. If this logical system is "sufficiently powerful" (the meaning of this is of course over our heads), then it will always be possible to formulate theorems which, within the framework of the system, cannot be proved or disproved.

At first glance it would seem that the trouble lies in a lack of axioms. That is not so. No matter how many axioms are taken, no matter how we supplement our system, there will always remain certain assertions about which nothing definite can be said.

After this marvellous theorem was proved, the whole problem of non-contradictoriness took on a different aspect.

We were silent on this point, just as we did not so much as touch on the totally unexpected *practical* application of mathematical logic. I mean computers.

We spoke in somewhat more detail about the second line of development that passes through Riemannian geometry to the general theory of relativity.

One more thing. The whole history of the development of non-Euclidean geometry appears as one of the most brilliant instances of unexpected turns in the history of science.

What appeared to be completely abstract, speculative, and theoretical meditations of mathematicians was in some marvellous way transmuted into things of extreme importance to practical physicists and even engineers.[4]

---

[4]For instance, the GPS system simply would not work without a proper account of the effects of special and general relativity!—A.S.

Chapter 12

Einstein

The essence and nature of any extraordinary talent are mysterious. That is a trite statement. But the bitter truth is that the mechanism, even the rough operating scheme, of that remarkable computing device that is our brain remains a mystery to science. We cannot make out how, in the brilliant scheme of evolution, nature fashioned some 14 to 17 thousand million elementary units called neurons into what is known as the human brain.

We do not even have a suitable answer to the question: "In what way does the human brain differ from that of some other animal?" We either confine ourselves to the general phenomenological reasonings of the biologist or to the scintillatingly clever but, alas, trivial paradoxes of the writer.

There is even less to be said about how the brain of a genius differs from that of a commonplace earth dweller. More, we do not even have any grounds to claim that there are some kind of organic differences of that nature.

It may very likely be that in every person some exceptional talent wastes away unbeknown to the world. It is a very enticing and consoling idea, and was developed at one time with the greatest pleasure by Mark Twain in his *Captain Stormfield's Visit to Heaven.*

The idea is, of course, suspicious. But there are no objective facts indicating its absurdity. It would perhaps be hard to find a better illustration of the level of our knowledge about the mechanism and biology of thinking. We hardly know anything and can only take note of the purely external characteristics of talent.

The oft-repeated phrase that "talent is work" defines one such characteristic. These words are commonly misunderstood to actually mean something; this is done all the more eagerly since the gifted, out of ostentatious modesty and with due respect for tradition, though at times quite sincerely underestimating themselves, point to work as the main source of their exceptional attainments.

Statements of this kind are many, but only a portion (and a small one at that!) is the truth.

Paganini claimed his wizard playing came from a supremely exhausting labour that enabled him to master the potentialities of his instrument.

He was wrong of course.

The writer Leo Tolstoy liked to say that his gift as a writer was not at all so great or significant, the truly important and valuable things being the moral ideas he preached that were so natural and simple.

I do not think that Tolstoy said what he thought.

Einstein, speaking of his genius, said a remarkable thing, and we shall come back to it again. But I think he had in mind something quite different and simply was compelled by circumstances (energetic newsmen) to throw a bone to the public.

So my idea is that we should not believe geniuses on this point. The eternal, mournful indignation of Pushkin's Salieri (a talented person, by the way) presents a better and more accurate picture of what a genius really is.

It is something incomprehensible.

When dealing with a normally endowed person, we can analyse and decipher a few things. We can then pick out, more or less clearly, techniques, experience, taste—all that comes as the reward of arduous exhausting labour.

For example one can almost always understand what is good and what is bad in Balzac's books.

But when you are imperceptibly charmed by the endless, rather clumsy and at times (horrible dictu!) simply grammatically incorrect phrases of Tolstoy; when you cease to watch the style, the techniques, the images and only follow the story of the piebald horse Holstomer, learning how he lived and died, and how many horses there were in the herd of his last owner.... When you can find dozens of more or less suitable explanations of why such and such a paragraph was written and what relative literary merits it has etc., but cannot grasp how it could have come to Tolstoy's mind to write that way and why you are left with that inexplicable conviction that that was precisely the way it should have been written...

Then we say that this is an anomaly, which can be recorded as such but cannot be accounted for.

The curious thing is that quite often a person who is a genius in one field is by no means a harmoniously endowed personality.

There are many confirmations of that. Not going too far, we can talk of Tolstoy again. Tolstoy the philosopher was a narrow-minded, biased and capricious personality.

We can bring this discussion of genius to a close by adding that the very conception of genius is extremely hazy and subjective, particularly when one deals with art, where objective criteria are still more nebulous.

In science too, ultimately, the deciding factors (or, to be more exact, their absence) are the same as in art, and that is why very often a first-magnitude star of today becomes noticeably faint tomorrow.

Some cases, incidentally, are unquestionable.

One is that of Albert Einstein.

As far as we can judge from reminiscences, the childhood years of Einstein did not in the least suggest that he would be an Einstein.

He was a quiet, reticent child. Usually children are full of life and energy, noisy, in a hurry, in a hurry to tell the world what they are.

But in every dozen there are one or two of the quiet kind. They do not take part in games and keep to themselves. They seem to be occupied more by their inner world than by the world around them. It may be that something has stirred up mistrust in their minds and they simply cautiously avoid people, instinctively believing that it is safer that way. Children of this kind are not liked in the rather merciless kingdom of childhood. They are continually being teased.

"Sissy", "mamma's boy", "weakling" are some of the international terms that often cause more anguish than, in later life, a rebuke by one's superior. At any rate, the mark they leave in the person's life is deeper.

Einstein was of the timid kind.

His relatives recall that he was called "mamma's boy" for his morbid love of the truth and fair play.

Another thing. He did not like soldiers. Neither the real ones marching along in bright new uniforms and helmets stamping in unison down the quiet streets of the towns of his Fatherland, nor the pretty tin soldiers that come in nice boxes. He did not like soldiers.

True, honesty and fair play are not so rare in children. The question, rather, lies in the age at which it ordinarily disappears.

Now as to this instinctive dislike of soldiers that is indeed strange.

There are not many boys like that, and one might suspect something out of the ordinary in such a child. But no, there does not seem to be the slightest indication that this "something" will, in fifteen years, flower into the theory of relativity.

There were other things that worried Einstein at this age.

I do not know whether the people around him noticed that at the age of ten or eleven this boy of well-to-do parents was going through a crucial internal drama, which in many ways determined the whole of his future life.

At least Einstein himself remembered; at the age of 67 he wrote:

"Even when I was a fairly precocious young man, the nothingness of the hopes and strivings which chases most men restlessly through life came to my consciousness with considerable vitality. Moreover, I soon discovered the cruelty of that chase, which in those years was much more carefully covered up by hypocrisy and glittering words than is the case today. By the mere existence of his stomach everyone was condemned to participate in that chase. Moreover, it was possible to satisfy the stomach by such participation, but not man in so far as he is a thinking and feeling being. As the first way out there was religion, which is implanted into every child by way of the traditional education-machine.

"Thus, I came—despite the fact that I was the son of entirely irreligious (Jewish) parents—to a deep religiosity, which, however, found an abrupt ending at the age of 12. Through the reading of popular scientific books I soon reached the conviction that much in the stories of the Bible could not be true. The consequence was a positively fanatic freethinking coupled with the impression that youth is intentionally being deceived by the state through lies; it was a crushing impression. Suspicion against every kind of authority grew out of this experience, a skeptical attitude towards the convictions which were alive in the social environment—an attitude which has never again left me, even though later on, because of a better insight into the causal connections, it lost some of its original poignancy."

This somewhat heavy passage demands more than a hasty reading. It is worth a most detailed analysis.

Note, firstly, that Einstein wrote his autobiography as a scientist striving to extract from his inner life with complete honesty only what deserves attention. He of course realized that this was no easy task at almost 70 years of age. He was even academically cautious in the title: *Autobiographical Notes*. Mostly he wrote about what he considered to be the only interesting thing in his life—the formation of his scientific outlook. His work.

There is no place for anything else in this self-obituary. There is no attempt to appear better, no ostentatious display of any kind. Actually it is a scientific paper. In every line one feels the desire to be as truthful and objective as possible in describing how he, Einstein, reasoned.

Such was the life of ten-year-old Einstein.

He did not like school. He recalled later that the teachers in elementary school seemed to him army sergeants; and the instructors in gymnasium were, to him, lieutenants.

Here we have the first riddle. One fairly often meets people who, irrespective of their culture and education, never reach the idea that a person needs something more than simple well-being. Some arrive at that conclusion at a mature age, or even at the end of their lives.

To one degree or another, this striving towards the mysterious "something else" is found in all children, but mostly in a very intuitive way of which they are not aware.

Einstein, on the contrary, reasoned with rigorous logic. As a result he arrived at religion, which was quite understandable, taking into account the conditions under which he lived.

So far there is nothing much out of the ordinary.

The amazing thing is that after reading a number of popular-science books the boy quite independently carried out a purely logical analysis and took a sharp turn away from religion, as a doctrine that is unsatisfactory. He even goes farther, arriving at a clear-cut conclusion of great social import: "...youth is intentionally being deceived by the state through lies..."

That was at the age of twelve.

And that was the conception that he carried with him throughout his life. If that is so, then wherein lies his, Albert Einstein's, "something"?

Very very cautiously, fearful of distorting the truth, he writes that partly consciously and partly subconsciously he came to the conclusion that for him life would be happy if he devoted himself to science. "The road to this paradise was not as comfortable and alluring as the road to the religious paradise; but it has proved itself as trustworthy, and I have never regretted having chosen it."

You can believe him, he was indeed one of the happiest people of our age. Perhaps he would have been just as happy even if we imagine that his work was not understood, not recognized and if he had to die an unknown eccentric engineer of the Swiss Patent Bureau at Bern, where, as a twenty-five-year-old youth he created the theory of relativity. Incidentally, at the end of his life he experienced something of this kind once again, in a sense.

Not in the sense of being famous, of course. He was the most recognized and most popular scientist in the world. He was almost as well known as Marilyn Monroe or the footballer Di Stefano. His name had become a symbol of the human intellect.

But physicists did not take much interest in the works written towards the end of his life. Yet it was only their opinion that carried any weight with Einstein.

Actually, not too much weight, because the decisive factor was always the opinion of Albert Einstein.

Why did he choose science?

Perhaps if a certain medical student had not suggested that he read popular-science literature he would have been a good musician instead of a brilliant physicist. Einstein played the violin from the age of six and was seriously and sincerely in love with music throughout his life. Then again he might have gone into inventing—another one of his passions. But such musings are idle.

Einstein himself, in later life, always said that if a person was born to be a physicist, if it was in his blood, then he would be a physicist no matter how his life turned out.

It's hard to say, he most likely was judging by himself. True, on one occasion, recalling his youth, he expressed the opposite view.

Be all this as it may, the existence of all the popular-science literature of the time would be justified by the single fact that it had some influence on the deeply thinking youngster of twelve who roamed the picturesque outskirts of the provincial Swabian town of Ulm in 1891.

Also soldiers' feet resounded on the streets of Ulm. They were the heirs of the victorious warriors of Moltke who twenty years before had routed France.

The military traditions of Ulm, it seems, went deeper still. In 1805—Ulm was then a first-class fortress—a wonderfully equipped Austrian army surrendered to Napoleon in a most scandalous fashion, virtually without fighting.

But, first of all, the army was Austrian, which means, formally speaking, not quite German, and this consequently implies "not at all German".

Secondly, the soldiers do not remember the defeats, for their heads are overfilled with victories.

Defeats were simply regrettable accidents, that's all.

So they marched.

That was probably when to the child Einstein came hate. A restrained, calm, somewhat cold and rational hatred. A hatred that invariably stayed with him his whole life. He could not stand militarism, war and slaughter. He viewed it all as the supreme concentration of human stupidity.

This became clear to him in his early years, and his view never changed.

The year was 1891. Fascism was a long way off. The crematoriums of Auschwitz and Maidanek were not yet built. They came later.

Germany was still to face the Schlieffen plan. The First World War. Marching armies. Exalted weeping women throwing flowers to their menfolk. Trainloads of soldiers. Ersatz food products. And the same women weeping, differently, over the endless stream of casualty telegrams from the Eastern and Western fronts. The final rout, the overthrow of the Kaiser, the Treaty of Versailles, inflation, ruin, hunger, and the epidemic of flu would all come later to the Germans. All these things would come before the Führer came.

True, there were a few things. For example, the bright uniforms and the Prussian general staff, antisemitism, and patriotic military marches, and fraternities, and—probably most important of all—an unquestioning reverence of titles.

Civilian or military, it makes no difference.

"Herr Privy Councillor! Oh! Indeed!.."

The great Olympian himself, Goethe (and a volume of Goethe could of course be found in every respectable family), even Goethe, ladies and gentlemen, was just as proud of his ministerial post in the miserable Weimar principality as, perhaps, he was of his poetry.

And Hegel? The great "Privy Councillor" Hegel, remember?[1] And his doctrine of the Prussian monarchy?

In short, the German state was consistent in its strivings to wipe out the very capability of independent (hence, critical) thought that is part and parcel of every normal human being, and to put in its place ready-made slogans, rules and traditions.

And they did a good job, one must admit. The system was polished to perfection by true craftsmen in the art.

The *Wacht am Rhein*, the sentimental "Lieder" of blue-eyed girls, and Wagner's operas, and gymnastics at school, the tales of ancient Nordic heroes at the history lesson, and traditional off-colour humour in cheap editions, and pedantic neatness instilled from early childhood, and absolute obedience to the head of the smallest unit of the state—the family. "Oh! Father said!.."

And, finally, the endless multiplicity and diversity of official, semi-official and non-official hierarchy of titles and ranks.

The hierarchy in the family, in the bureaucracy, in the military service; the hierarchy of numberless Vereins, fraternities, unions in sports, unions at the factory, in music, the arts, the sciences, in literature and religion; unions of lovers of hunting, lovers of song birds, unions of beekeepers, yachtsmen, and so on and on.

All this created and cherished a conventionality that was both self-satisfied and humble; it created people that forgot that they were capable of thinking, people for whom a dictatorship appeared to be the most natural form of power imaginable for the reason that each in himself was a dictator

---

[1] Hegel was not himself a Privy Councillor. One of his students was.—A.S.

on a microscale.

The remarkable thing about the infinitely poisonous character of this whole demoniacal machine was that it fed on very decent feelings and aspirations—patriotism, respect for one's elders, sports...

But what of the people themselves?

I would say that whether at the beginning of the fifteenth century or the beginning of the twentieth, or even during the years of fascism, the German people did not differ in any way from any other people. There can be no question that a hundred thousand scoundrels can be found in any large country. The historical situation in Germany at the end of the 1920's was such that precisely this group came to power. Possibly, accidental circumstances played an appreciable role here.

True, the prerequisites for this accident were already prepared. Incidentally, I would not be saying anything original or new if I added that roughly the same prerequisites were available in any of the large imperialist states.

This had been noticed many times. One can mention the science fiction novels of such writers as Sinclair Lewis and H. G. Wells, where the picture of fascism developing in the United States or in England was depicted rather convincingly. Perhaps the greatest danger of the demagogy of fascism lies in the fact that it is not new or exceptional in any way.

If fascism is a disease of the human race, it is an ancient affliction. States of the fascist type existed in all ages. Egypt, Sparta, Rome—all these ancient regimes preached just about the same ideology as the Nazis. So Hitler did not have to concoct anything particularly new. True, he added a goodly portion of social demagogy, which Egypt got along without but which ancient Rome already found it necessary to include.

And, of course, one of the basic axioms of the system was nationalism.

Nothing very original about that either. From time immemorial, flattery, even of the crudest kind, has always been excellent bait to lure the hearts of members of the human race. It is always nice to hear that you are better than the next man. All the more so, when you yourself are not so sure of the fact.

Now if the flattery is kept up insistently enough, one begins to believe.

Every empire-building state since the pharaohs of Egypt has brought nationalism into play as a means of attracting and uniting the people.

The idea is simple and naive, a truism.

The emperors of Rome, Genghis Khan, Napoleon, Hitler have all employed the same technique, just as tried and tested as complimenting the woman one wants to seduce. Towards the end of his life, Einstein gloomily remarked that people learn but little from the lessons of history because each new act of stupidity appears to them in a fresh light.

That this system produces results even in our "enlightened" age was, unfortunately, demonstrated in the Second World War. But we must repeat: the fact that most of the German people accepted fascism in one form or another does not, of course, imply that the Germans as such are less responsive to the generally accepted moral norms than the Russians or the French.

Now if the question of the responsibility of the German people as a whole for the rise of fascism comes up, then with just as much justification the question could be addressed to all the capitalist states of our planet, which with comparative calm watched Hitler advance from the Beer Hall putsch in Bavaria to the furnaces of the concentration camps and the mass shootings in Russia, Poland, Yugoslavia...

Today, twenty odd years after the end of the war, today when it is possible to judge with relative objectivity, one should hardly throw all the horrible blame onto the German people.

All the more so since that nation too paid a sufficiently dear price. Among the victims of the Nazis were also those Berlin youngsters who during the last April days of 1945, crying from sheer fright, went at Russian tanks with Faust-Patronen sincerely believing that they were fighting and dying for their Fatherland.

These arguments are probably just as true as the fact that the active SS men and the "creative" and initiative Hitlerites of the punitive expeditions and death camps should be judged and exterminated today, twenty and more years after the war's end; they should be shot calmly and with a clear conscience, "without anger and bias", on the basis of the very same reasoning that professional murderers and recidivists are wiped out.

We may recall that once upon a time one of them got the idea of writing "Jedem das seine" on the gates to the Buchenwald concentration camp—to each his due.

Why do I write this here? Because that was approximately the way Einstein thought. He hated fascism his whole life.

Humanism and the all-permeating kindness that was Einstein's does not appear to link up with sentimental all-forgivingness, which, as a rule, stems from indifference and gets along very well with stone-cold egotism.

This is nicely and precisely described by Leopold Infeld in his recollections.

Unfortunately, in recollections and biographies, Einstein very often appears a kind of eccentric emanating an endless stream of gentleness and far removed from any thought that there can be meanness, deceit and wickedness in the ordinary day-to-day world. Such writings are mostly irritating, for, whether intentionally or not, the authors make Einstein out to be stupid.

We will do better to quote Infeld:

"I learned much from Einstein in the realm of physics. But what I value most is what I was taught by my contact with him in the human rather than the scientific domain. Einstein is the kindest, most understanding and helpful man in the world. But this somewhat commonplace statement requires comments.

"The feeling of pity is one of the sources of human kindness. Pity for the fate of our fellow-men, for the misery around us, for the suffering of human beings, stirs our emotions by the resonance of sympathy. Our own attachments to life and people, the ties which bind us to the outside world, awaken our emotional response to the struggle and suffering outside ourselves.

"But there is also another entirely different source of human kindness. It is the detached feeling of duty based on aloof, clear reasoning. Good, clear thinking leads to kindness and loyalty because this is what makes life simpler, fuller, richer, diminishes friction and unhappiness in our environment and therefore also in our lives.

"A sound social attitude, helpfulness, friendliness, kindness, may come from both these different sources; to express it anatomically, from heart and brain. As the years passed I learned to value more and more the second kind of decency that arises from clear thinking. Too often I have seen how emotions unsupported by clear thought are useless if not destructive."

I am only sorry that I did not write this passage myself. Without sentimentality and passion, without melodrama and tragedy and a soul-divesting self-analysis, with the calm logic of the physicist, Infeld has here formulated the best towards which every person strives.

But, as we know, the road to hell is paved with good intentions. To strive does not mean to accomplish.

That is not all. Infeld writes further.

"Here again, as I see it, Einstein represents a limiting case. I had never encountered so much kindness that was so completely detached. Though only scientific ideas and physics really matter to Einstein, he has never refused to help when he felt that his help was needed and could be effective. He wrote thousands of letters of recommendation, gave advice to hundreds. For hours he talked with a crank because the family had written that Einstein was the only one who could cure him. Einstein is kind, smiling, understanding, talkative with people whom he meets, waiting patiently for the moment when he will be left alone to return to his work."

It is hard to believe that Einstein found any pleasure in talking to this psychically unbalanced man. It would be just as naive to think that Einstein hoped, through such an encounter, to heal the man. But he probably believed, after weighing and analysing the case, that he might bring about a slight and temporary improvement in the state of the patient, and thus alleviate the life of the family. It was with this purely hypothetical possibility in mind that he considered it necessary to tear himself away from his work—his only god.

And when Einstein arrived, inwardly, at some conclusion, he did not leave it as some speculative dogma; for him, thought signified, above all, action coordinated with thought.

Here I am writing something in the nature of a biography, yet all the time I hear Einstein's own calm remark in his *Autobiographical Notes* that he himself could not hope to convey exactly his own thoughts and his inner world. Quite naturally, a biographer would succeed even less.

Even when one is dealing with a rather common personage, this is an insurmountable problem. It becomes absolutely unresolvable when one attempts to write about a man the stature of Einstein.

The more so that Einstein's own writings on this matter are naturally very often contradictory, while a biographer's writings are unavoidably subjective.

Yet in the case of Einstein, it appears, paradoxically, that some things are simpler than even in the biographies of some of the long since forgotten "immortals" of the French Academy of Sciences.

This may be due, again, to the fact that in his emotional life too he followed with purely German pedantism the clear-cut logical criteria of a consistent and realistic humanist that he had worked out for himself in his childhood and early youth.

It was more difficult to shake his convictions here than in his general theory of relativity, though he himself did not in the least overestimate his virtues.

The calm and saddening skepticism of a mild, clever and kind scholar made completely and unconditionally impossible any sort—so common in people of that character—of narrow-minded self-satisfaction of the righteous man who has learned the truth and is communicating it to the lost world.

Shortly before his death he wrote to Max Born that what every man has to do is to be a model of purity and have the courage gravely to maintain ethic convictions in a society of cynics. He added that he had strived for a long time to act in this manner and had succeeded—to some extent.

These sad and wearied words were spoken by a man who was—everyone said—always charged with a natural inner and invincible cheerfulness. They indicate that Einstein perpetually felt a heavy inner awkwardness through-out his conscious life. He was constantly worried of being too speculative and too passive in the struggle against baseness and absurdity that stood out so conspicuously in the surrounding world. Above all, he was oppressed by the obvious absurdity of what was happening in the world.

How it was and why Einstein decided that social activity was not his business, I do not know. Perhaps he did not see any real ways out. It may be that emotions and feelings were decisive. To some extent unconsciously, obeying the exhortations of his heart, he found his identity in physics.

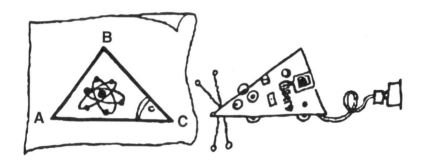

Perhaps some role was played by the inner reticence and individuality of his thinking. And after the choice was made, all else was pushed aside and into the background by the chief passion of his life.

But the surrounding world was never for a moment switched out of his mind. In actual life he was constantly encountering political intrigues, malice, and human passions—he could not stand aside from these things, for he clearly and firmly realized that a human being has no right to do so.

This idea is simply a repetition of what was said earlier, where we also mentioned the "German pedantry" of Einstein. Well, it needs hardly be said that by pedantry here was meant a kind of integrity and ultimate logicality of character. Since these features are commonly thought to be intrinsic to the national character of Germans, I used the adjective "German".

In the Search for Beauty

I do not intend to take it back because—and I believe it is well worth saying—Albert Einstein, though born a Jew, had an American passport, and was a consistent and unconditional internationalist in his convictions, an internationalist in both mind and heart, still was all his life a German, a German in language, in culture, in customs and in the hardly perceivable habits, eccentricities and minutiae that, ultimately, go to make up a nation, patriotism, and love of one's motherland.

He was a German in his rather heavy (particularly in his youth) academically dry humour. In later years, the ponderousness retreated and his pronouncements became polished and aphoristic, though again this was the humour of Heine rather than that of Twain or Bulgakov.

He was also a German in his somewhat contemplative love of quiet nature and walking in the countryside, in his household habits, in his passion for Mozart, his penchant for analysing philosophical problems, and his love of his mother tongue.

The last words he pronounced were in the language of his childhood, German, and they were not understood by the nurse who was the only one with him when he passed away.

After twenty years of life in America—it is hard to imagine anything more paradoxical—he was just reaching the point (said one historian of physics) where he could handle the English language satisfactorily.

But even during the latter years of his life he preferred to speak German if his companion spoke that language.

He was homesick just like any ordinary burgher who might have come to the United States on business and settled down for the rest of his life. For—and this incidentally was the credo of Einstein himself—there are things and concepts common to all people irrespective of their intellect and culture. Now in matters of ethics, in the aspect of norms of human conduct, Einstein was a fully convicted democrat who recognized both in word and in deed the complete a priori equality of human beings.

I feel I must stray again and relate a story, which, though it almost sounds like a joke, gives a very accurate picture of Einstein's stand and style in his dealings with people.

There seemed to be a vacancy open at a certain institution, and four different applicants came to him one after the other for letters of recommendation. Einstein gave letters to all four.

To the surprised questions of his friends he replied calmly that he saw nothing strange or extravagant in what he had done, for in each case he gave different reasons for his choice of candidate and it was, he said, up to the employer to do the choosing.

Let us return to 1891, to the town of Ulm and to the twelve-year-old boy who was experiencing a wonder. It was contained in a book on Euclidean plane geometry. Euclid was a revelation to Einstein, and it remained so to the end of his life. Shortly before his death he said words to the effect that if Euclid's work could not fire one's enthusiasm in youth, then that person was not born to be a theoretician.

Einstein's recollection of this wonder on the fourth or fifth page of his *Autobiographical Notes* is just about the last purely autobiographical recollection.

A few words follow about his education at the Polytechnic Institute of Zürich, then just in passing a remark or two about the system of instruction and roughly fifty pages of Einstein's ideas concerning modes of thinking, epistemology and, of course, physics, as always.

But one should not get the idea that this way of constructing an autobiography is another one of those cute absurd absent-minded ways of the aschetic monk. Don't ever try to represent Albert Einstein as a kind of Jacques Paganel of physics.

A few pages later he gives a clear-cut and calm explanation of his somewhat extravagant manner of presenting things.

"And this is an obituary?" asks the puzzled reader. I feel like answering: "Why yes, of course. Because the most important thing in the life of a man of my make-up is what he thinks and how he thinks and not what he does or experiences."

That is why Einstein recalls the wonder of geometry and does not even mention his Nobel Prize.

This idea of a "wonder" as of something that the human mind encounters that contradicts all established notions, is very persistently repeated by Einstein throughout his life.

In reply to a reporter's question as to how it happened that Einstein and not somebody else discovered the special theory of relativity, Einstein remarked that he was rather late in developing mentally and that for this reason he still retained the perception of a child at the age of 20–25. And so when, unencumbered, he meditated on the situation of things in physics, he naturally was surprised like any normal child would be, but since he was at that time twenty something years of age, his intellect was more developed (this he admitted) than that of a normal ten-year-old boy and so he was able to obtain results that comprised the special theory of relativity.

Penetrating to the kernel of matters here, we find that there is an important and very essential idea behind it all, that the scientist should constantly experience a feeling of wonderment and regard all the phenomena of nature in an unprejudiced manner; he should reject all dogmas and authorities... In short, he should think and not quote. True, this was not an original thought. Plato had already put the idea neatly when he said: "Wonder is the mother of science."

Today this is such a truism that no self-respecting writer risks repeating it, yet there is no second Einstein. Obviously, there must be something more. But, sad as it is to admit, we are rather in the position of a eunuch being told the meaning of love.

So young Einstein experienced one wonder after another. Between the ages of twelve and sixteen he discovered mathematics, and the purely emotional impression that this new world, the world of precise logic and unbridled imagination made on him, was exceptional.

At about this time Einstein experienced yet another wonder, purely psychological.

"The fact that I neglected mathematics to a certain extent had its cause not merely in my stronger interest in the natural sciences than in mathematics but also in the following strange experience. I saw that mathematics was split up into numerous specialities, each of which could easily absorb the short lifetime granted to us... my intuition was not strong enough in the field of mathematics in order to differentiate clearly the fundamentally important, that which is really basic, from the rest of the more or less dispensable erudition. Beyond this, however, my interest in the knowledge of nature was also unqualifiedly stronger; and it was not clear to me as a student that the approach to a more profound knowledge of the basic principles of physics is tied up with the most intricate mathematical methods. This dawned upon me only gradually after years of independent scientific work. True enough, physics also was divided into separate fields, each of which was capable of devouring a short lifetime of work without having satisfied the hunger for deeper knowledge. The mass of insufficiently connected experimental data was overwhelming here also. In this field, however, I soon learned to scent out that which was able to lead to fundamentals and to turn aside from everything else, from the multitude of things which clutter up the mind and divert it from the essential."

This is amazing. It is not so important whether Einstein, at the age of sixteen to twenty had formulated to himself these things or whether the decision was to some extent not consciously felt in his own mind.

•

The amazing thing is the maturity of such a choice. Such lucid critical thinking in general is very rare, and is something practically unheard of at the age of sixteen or eighteen years.

Indeed, take a look at what we have. Here is a young boy of sixteen carried away by mathematics. The integral, the fundamentals of analytic geometry are a great source of pleasure, of such joy that nothing else can compare. He of course realizes that he is gifted and that his talent stands out on the general background.

He had every possibility of a free choice (and this is most essential), no circumstances of life compelled him. Even more, if one takes into account the purely external influences, then there were more points in favour of mathematics. The Polytechnic Institute of Zürich had a number of brilliant mathematicians such as Minkowski. There were no outstanding physicists though. Einstein himself said later that up to the age of thirty he had never seen a real theoretical physicist.

Given starting conditions like these, it is hardly possible to conceive of a young man giving up mathematics for the cognate subject of theoretical physics.

A change over to poetry or, say, music would have been, psychologically speaking, more understandable.

I feel that the problem was resolved by an amazing feature of Einstein's character, which, obviously, was already fully mature in those years, and that is a total absence of intellectual conceit that is so natural among gifted young people.

He always appraised both his potentialities and his results soberly and calmly. He never played at ostentatious modesty and he knew—he said it openly—that his works represent the greatest result of twentieth-century science.

At the same time he knew (or he thought he knew) that he would not become an outstanding mathematician.

And so he gave up mathematics.

Throughout his lifetime, Einstein's relationships with mathematics were rather complicated. On the one hand, in later life, he time and again regretted his youthful self-confident conclusion that physics required only the fundamentals of mathematics and that the more sophisticated matters could be left to professional mathematicians. He became convinced of this error when he began working on the general theory of relativity. During the first stages, he had to ask the help of his friend Marcel Grossmann in the mathematical portion.

In later years, Einstein's views changed. His main works—at least outwardly—are works of a mathematician.

Nevertheless, he always remained a physicist in mode of thought and in his approach to problems.

I shall not risk getting into a discussion about the similarities and differences of the theoretical physicist and the pure mathematician. Suffice it to say that there is a difference. And a rather essential one, as witness the following amusing exchange of wit between Einstein and Hilbert.

In 1915, Hilbert took a liking to the theory of relativity and decided to try his hand at physics believing that substantial progress would not be made without mathematicians.

As he rather cleverly put it without excessive modesty, "physics is actually too difficult for the physicist". And his work, being naturally at the ultimate mathematical level, somewhat lacked the physical content.

In a letter to Ehrenfest, Einstein rather spitefully replied for the physicists when he described Hilbert's work as the tricks of a superman. Towards the end of his life, Einstein remarked to the effect that mathematics is the only perfect way of leading yourself around by the nose.

We will not attempt to draw any moral here, but will simply repeat that no matter how mathematical Einstein's works were, he always remained a physicist.

It is now time for us to note one important factor. Though Einstein repeatedly said that the response of the community—recognition on the part of his colleagues—was extremely important to him, and this was of course true, his own appraisal of his work was the decisive factor.

To the very end of his days he could not reconcile himself to the basic ideas of quantum mechanics (which he relegated to the class of ephemeral physics) and though he remained alone he never changed his opinion.

In the same way, he was the only physicist in the world who, without any external prerequisites and after having earned fame and recognition, worked for ten years (between 1905 and 1915) on the problem of the gravitational field.

Standing quite outside the range of interests of the physicists of that period, he created the general theory of relativity.

Perhaps due to a variety of accidental circumstances he became the most famous scientist in the world. Calmly and somewhat sardonically he withstood a virtual avalanche of honorary awards, medals and distinctions (including the title and attire of honorary chief of an Indian tribe).

And then for another 35 years he worked intensely on possible extensions of the general theory of relativity—the unified field theories. He was doing so remaining practically alone, actually without any recognition or moral support and appearing in the eyes of the new generation of quite self-confident theoreticians of the 1930's to 1950's something in the nature of an aging monument.

Incidentally, he once mentioned to his wife that the results he obtained in the forties were the biggest contribution that he had ever made.

Who knows whether he was right, as he almost always was when the subject matter was physics? The only thing to be said is that one can see now the increasing interest in the general theory of relativity and, in particular, in the investigations of Einstein carried out during the last years of his life.

But perhaps that too is just a fad which physicists are prone to follow like women do fashions. Or it may simply be an expression of a certain disappointment, a crisis in modern theoretical physics.

Yet perhaps the foundations of the physics of the future are indeed to be sought in Einstein's works on the unified field theory.[2] At any rate, the scientific career of Einstein, beginning from his general theory of relativity, is an unparalleled anomaly in the history of science.

And if one speaks of the purely personal aspect of the matter, this story amazes us and causes more respect than the purely mathematical giftedness of Einstein, which after all is not related directly to his human qualities.

In passing let us add that on the side (even if we count from 1920 onwards), Einstein carried out a range of researches totally unconnected with relativity theory, but, if one splits them between several scientists, the latter would have good chance to fill five or six vacancies at an election to the Academy of Sciences.

We may again add that his results in the theory of Brownian motion and the photoelectric effect (this was in 1905) were in themselves sufficient to have ensured the author an exceptional place in the history of physics.

We might also recall that the most fashionable and promising trend today in quantum statistics has as its source the theory of the heat capacity

---

[2] Einstein's *idea* of unified theory turned out to be very deep and fruitful. In particular, we know today that the *electromagnetic* and the so-called *weak* interactions have common origin and are described by a unified theory. But Einstein wanted to discover the unified theory of gravitational and electromagnetic interactions. His attempts in this direction were not really successful, this problem turned out to be very difficult and it is still not solved today. But most physicists now believe that it *has* a solution, and the unified "Theory of Everything" describing *all* fundamental interactions, including gravity, does exist.—A.S.

of crystals, which just by the way was proposed by Einstein in 1908.

Finally, Einstein's rejection of quantum mechanics, his paradoxes, yielded so much material for an elucidation of the fundamentals of that field that in themselves they can be considered first-magnitude works of science. Then, too, he obtained a number of very important results after 1915 in various parts of the quantum theory.

But for him all of these were only a mental game and a pleasant recreation from the main thing—the unified field theory.

So we have Einstein at the Zürich Polytechnic Institute majoring in physics and neglecting mathematics. He even skipped lectures—not to spend his time idly but the better to utilize it. Before arriving at Zürich together with his family, he had already visited Milan and had experienced a number of small unpleasantnesses—he was told to leave the gymnasium at Munich for unhealthy skepticism; he failed once in an examination in zoology and botany at the Polytechnic Institute.

But these events, which for another person might have played a decisive role, were for Einstein merely unpleasant trivia.

The die was cast, and his natural bubbling-over cheerfulness and clear-thinking head dismissed all these and other bumps and scrapes that came his way. He wrote that he was never in a gloomy mood unless he had a stomachache....

Judging by his letters and the recollections of relatives, Einstein at 20–25 years of age was a strong life-loving young man with a passion for music, painting, literature, hiking, with a gift for the joke, though, honestly speaking, his humour was not always up to the mark. He was a bit extravagant, a trifle forgetful (like forgetting the keys to his flat after his wedding or using a doily for a scarf). But this was all natural, for it stemmed from a striving towards greater inner freedom, though—and this is important—there was never any hint of this constant urge for inner independence ever building up into egotism and a disregard for those about him. This was precluded by an inborn culture and a consciously developed mildness.

In a word, he was a nice well-mannered young man, broad-minded, without a trace of conceit or morbid reflections. One could readily foresee his future as a school principal or a top-class expert in the patent bureau, where at that time he was only rated third-class. One could see him a great lover of music and literature, reading Sophocles, Racine, Cervantes, discussing the treatises of Spinoza and Hume, which he was then reading with a group of friends. One could picture Einstein on a mountain hike animately discussing Mozart, Alexander of Macedonia, Aeschylus, Beethoven, Kant, Archimedes, Cleopatra, Newton, Cuvier, Confucius, Anatole France....

Later, we might see him the author of progressive articles on the history of science, or music or pedagogy....

In short, his letters and the recollections of people who knew him draw a picture of a very nice young man disturbingly ordinary.

One finds it hard to believe, then, that this was Einstein and not just some pleasant, educated well-mannered, clever young man.

Perhaps there is one thing, Einstein's ability to dispense with all externals when the discussion turns to philosophy or physics. But no, this was not a very exceptional feature among the young people of those days.

Actually, however, an explosion was in the making.

And it came in 1905.

I must repeat that any one of three works of Einstein that appeared in that year—the theory of Brownian motion, the theory of the photoelectric effect, and the theory of relativity—would elevate the author to the rank of extra-class theoretician.

It remains a psychological mystery whether Einstein himself fully realized what he had accomplished.

If he did—and everything about Einstein and his later pronouncements on this score suggests that that was the case—then we must admit that intellectually he must have been very much alone, and the pleasant people about him did not even notice anything out of the ordinary, while Einstein himself, tactfully reticent, tried not to suppress his friends whom he liked in a very human way. Otherwise how are we to explain his letter to Habicht, one of his friends of the Bern period?

This unique epistle begins: "Dear Habicht, the silence between us is sacred and the fact that I am interrupting it with mere twaddle may seem a profanation." And so on in Einstein's old-fashioned ponderous playful style, calling Habicht a "frozen whale" and fancifully upbraiding him for not sending his dissertation, which Einstein was eager to get and read "with pleasure and interest".

But the best joke of all, one quite worthy of Heinrich Heine, is hidden at the very beginning of the letter, because what is being offered as mere twaddle is the following:

"In return [for Habicht's dissertation] I promise you four papers, the first of which I will send soon because I am expecting the author's copies.

"It is devoted to radiation and light energy and is very revolutionary, as you yourself will see, if you first send me your work.

"The second paper contains a determination of the true size of atoms by means of studying diffusion and internal friction in liquid solutions.

"The third demonstrates that in accordance with the molecular theory of heat, particles of the order of 10^{-3} mm suspended in a liquid experience apparent chaotic motion due to the thermal motion of the molecules. Biologists have already observed such motions of suspended particles; their term is Brownian molecular motion.

"The fourth paper is based on the electrodynamics of moving bodies and modifies the conception of space and time: you will be interested in the purely kinematic part of the work..."

Habicht certainly did not lose out in this exchange.

Definitely, there were some scenic elements in this letter. It is difficult, however, to evaluate their exact proportion.

On the one hand, the emphatic modesty with which Einstein addresses Habicht looks as a play or else a traditional courtesy. It is hard to take seriously the rather timidly expressed hope that in a paper where, in passing as it were, our conceptions of time and space are overthrown, there might be something of interest to Habicht. We would then get a picture of Einstein verging on that of the village simpleton.

Yet on the other hand—and this is evident from all future letters, from Einstein's whole life—there is the sincere awareness, confidence, conviction (what have you) that Habicht is a man, a personality and has the same value as he, Albert Einstein, and is not different before any law. Above all, before the inner law that Einstein obeyed in his youth, in maturity and in old age.

Most likely the impression of a certain ordinariness in the person of

Einstein (I speak purposely of his youth when his associates and companions could not yet know that they were dealing with the greatest physicist in the world) was largely due to Einstein's overriding feeling of democracy, and an egalitarianism just as natural to him as his desire to study theoretical physics.

I have already spoken of this, but I want to repeat because to people of the twentieth century this trait of an outstanding person is probably the most cherished; one is especially attracted to a man who, when placed in an exceptional situation either due to his own merits or to a more or less accidental set of circumstances, remains democratic and humanistic not only in form but in essence too.

And note that for a scientist of Einstein's stature, there were not less but perhaps more grounds and conditions to become, at least in the community of his associates and pupils, a more unbridled and cruel dictator in the sphere of the intellect than any actual dictator has in the sphere of political life.

Self-confidence, which expands into capriciousness, intolerance, and conceit, unfortunately often attends outstanding (as well as mediocre) scientists, who in this respect only fall short of poets and prima donnas.

Such things are not usually written in books and memoirs yet that is the case.

True, I also can judge Einstein only on the basis of the books and memoirs, but this case appears to be absolutely clear. Einstein did not have a single one of these traits to even the slightest degree.

That is yet another psychological enigma associated with the name of Albert Einstein, and by far not the last in significance.

Einstein stood the test of fame in just as easy-going a fashion—hardly noticing it—as he did his failure at the exams at the Polytechnic Institute of Zürich.

That, approximately, is the picture I have of Einstein.

One thing remains. It is very important. It is the attitude of Einstein to violence and war.

Willy-nilly, from about the 1920's onwards, when he had become world-famous, and the nationalistic, antisemitic fascist scum of Germany had begun victimizing him and his works, to the end of his life he was closely associated with political affairs at large.

One cannot say that he tried to evade burning political issues of the day. He clearly realized that, firstly, such a thing was simply impossible (whether he liked it or not is a different question), and secondly, he felt that he simply had to interfere wherever he believed that some good could result.

But here he found himself in a sphere where, from his point of view, very many things were unpredictable, uncontrollable, and unexplainable.

Because Einstein was extremely perceptive, he could probably picture to himself and account for the psychology of officers of the Prussian general staff, but to conceive of a human being reasoning and acting like the commandants of extermination camps, like the men in punitive expeditions and the hundreds and hundreds of thousands of SS men, or to understand how it came about that the leaders of quite a few countries could be morally and intellectually about on a level with those very same SS men was something beyond the capacity of Einstein. This was because he unwittingly overestimated the human intellect.

In the 1930's he who was a convinced and consistent pacifist had to say "now is not the time for pacifist ideas", for (this was a natural, immediate conclusion) the only way to halt the spread of fascism is by use of military force.

In what followed he was a witness to an involved, stupid and dirty political game. He saw politicians of the twentieth century adhering to the old-fashioned, naive criteria of humanitarianism to almost the same degree as Genghis Khan. He witnessed the Second World War, and he saw events after the war build up into a fresh threat of yet another war. He was to some extent responsible for the making of the atomic bomb, for he had written his famous letter to Roosevelt.

In reminiscences of Einstein, writers often speak of the so-called "Einsteinian tragedy of the atomic bomb".

To my mind, it was not the bomb.

From the standpoint of reason and logic (and these factors were always decisive for Einstein) he was irreproachable.

He wrote the letter in August, 1939, when there was a direct and immediate danger of Hitler making the bomb and when the only reasonable solution was to get it before fascism did.

He fully realized that he had had nothing to do with the cold-blooded senseless murder of tens of thousands of Japanese in Hiroshima and Nagasaki, all the more so since in 1945 he wrote Roosevelt asking him not to allow the military use of the bomb.

To Einstein, the atomic bombardment of these cities was in the way of the last act of human barbarism, final proof of the hopeless position of the scientist, the absurdity of the social structure, the unconditional abnormality of human beings in seats of government.

Of course, this gloomy conclusion was aggravated by the purely emotional realization that he, Albert Einstein, was connected with the explosion, however indirectly. But this was only an incidental factor. More depressing still was the fact that during those years he at times lost faith in the possibility of any social and moral progress, yet this ran counter to everything Einstein stood for. However, here too he remained true to himself, to his manner of outwardly dispassionate, calm analysis.

He learned of the explosion by radio. It was announced by Truman: "We have spent two billion dollars on the greatest scientific gamble in history and won."

Einstein's first reaction was one of grief and despondency. Yet he realized that the tragedy had nothing to do with the discovery of the chain reaction. He wrote, "The discovery of the fission of uranium does not represent a threat to civilization any more than the discovery of matches does. The future development of humanity depends on its moral code and not on the level of technology." The same idea was expressed further: "The world was on the verge of a crisis, the whole significance of which was not perceived by those who have the power to decide between good and evil, that the newly released atomic energy had changed everything, leaving unchanged only our mode of thinking...

"The solution of this problem lies in the hearts of the people."

But the fact that he saw all this so clearly did not make things easier. Towards the end of his life his supply of natural cheer was running out, and his depressed state of mind only aggravated the mercilessly critical view he took in appraising himself and his work.

"There is not a single idea which I am convinced will stand the test of time. At times I am in doubt about the correctness of the path I have taken. My contemporaries see in me at once a rebel and a reactionary who, to put it figuratively, has outlived himself. That is of course a passing fad caused by nearsightedness. The feeling of dissatisfaction comes, however, from within."

Einstein's seventieth anniversary was being celebrated when he wrote this letter to an old friend. Honours never moved him, and now still less. He sadly concluded: "The best that life has given me is a few real friends, bright and cordial, who understand one another like you and me."

One year before his death, when he declined the invitation to be present at the fiftieth anniversary of the creation of the special theory of relativity, he wrote in the same spirit:

"Old age and illness do not permit me to take part in such ceremonies. And I must admit that in part I am grateful to fate, for everything that is in the least associated with the cult of the personality has always been a torture to me.... In my long life I have come to understand that we are a great deal farther away from a real understanding of the processes of nature than most people today realize."

There may have been more optimistic notes at other moments, but in general his mood was melancholic. Nevertheless he continued to work. His cheer at times left him, but never his clear analytical mind, which functioned flawlessly to the end. He never changed his views or convictions in the least. They merely took on more sombre tones.

As before he was always ready to respond to a letter or to define his ideals, though more often one would hear him say, "people have gone mad", "the world is on the brink of a catastrophe". During these years of the "cold war" the situation in the United States was grave. At such times extremists always come to the surface. The notorious Anti-American Activities Committee was in work. The slightest deviation from official political views was dangerous. Naturally, the intellectuals—the most wide-awake portion of the nation—were first to come under suspicion.

In Einstein's letters and speeches of this period, one sees more and more a bitter yet courageous stoicism. Not a drop of sentimentality.

As before he was very far away from any kind of complacent all-forgivingness.

In reply to an American teacher, he wrote:

"Frankly, I can only see the revolutionary way of noncooperation in the sense of Gandhi's. Every intellectual... must be prepared for jail and economic ruin, in short, for the sacrifice of his personal welfare in the interest of the cultural welfare of his country.

"If enough people are ready to take this grave step they will be successful. If not, then the intellectuals of this country deserve nothing better than the slavery which is intended for them."

Is not this the same as saying "Yes, the people of a country deserve the government that they have"?

He continued to receive letters and no matter what he thought, what his mood was, he considered it his duty to help by writing to those who felt they needed his aid. His work suffered, of course, but what was there to do.

As he puts it with a bit of irony just a year before his death, "The time I need for meditation and work I have to steal like a professional thief".

And yet for all that—whether he was disappointed in humanity or in the level of human knowledge—he continued to work to the end.

Now I can see how poorly I have succeeded in writing about Einstein. To say nothing of other things, I realize that Einstein appears here unreal, unbelievable, too good.

But that was what he was.

Perhaps his greatest weakness was a somewhat cruel irony. He acutely saw the weak sides of people and at times he overindulged in his humour. Of course, he was no saint and would get irritated over purely personal matters. And probably at times unnecessarily so. Particularly in his youth.

He was not ashamed of writing very bad poetry, even liked to, and he would send his verses to his friends. He gave concerts eagerly though his violin playing was far from brilliant.

Finally,—true, this is only a suspicion I have, based on circumstantial evidence—I think he was inclined to courting ladies in a rather old-fashioned sort of way. That would seem to complete the list of his sins.

His most salient trait was that in his private life he strictly adhered to those beautiful principles that he espoused publicly. People of this kind are rare and the more so the higher their standing.

Naturally, a man is best tested in the face of death. From 1948 onwards Einstein knew that at any moment his life might end suddenly. He had said a number of times that he was not afraid of death, that the expectation of death would not change anything in his life, and now he proved it.

It did not, except perhaps his diet which he tried to observe. Just as thirty years earlier, he was calmly sarcastic when speaking about his possible departure to a better world and, when in April 1955 his time came, he remained the way he had always been.

Einstein suffered greatly, and he knew that he would die. But whenever there was any improvement he reverted to his beloved irony and stoically awaited events. He died in his sleep.

Einstein was probably one of the most likeable persons in the history of humankind.

Name Index

Printed in the United States
By Bookmasters